EIA
the environmental impact assessment process
WHAT IT IS and WHAT IT MEANS TO YOU

A Manual for Everyone Concerned about the Environment and Decisions Made about its Development

DIANE WIESNER, Ph.D., B.Sc. (Hons), M.A. Scientist
and Environmental Consultant

PRISM PRESS

Published in 1995 by
PRISM PRESS
Bridport,
Dorset DT6 3NQ
Great Britain.

Distributed in the USA by
THE ATRIUM PUBLISHERS GROUP
3356 Coffey Lane
Santa Rosa
CA 95403

ISBN 1 85327 093 8

Made and printed in Great Britain by the
Guernsey Press Co. Ltd., Guernsey, Channel Islands.

CONTENTS

INTRODUCTION

In less complicated times, a study of the environment and human impacts upon it focused mainly on climatic conditions and the way they affected human beings. Recommendations on ways to overcome or manage the worst aspects were raised.

Humans themselves create major problems through their normal daily activities and their struggle to survive. In our more complex life today with mechanisation, robotics, vast amounts of scientific data and computer technology, many more factors have relevance than was the case merely thirty years ago.

Too often in developed westernised nations, the assumption is made that human activities, major projects and changes to the existing environment will inevitably bring more problems and wreak long-term damage on what is a fragile planet where the continuance of life is precariously balanced.

How have we come to this view? Why are studies on the impacts and effects of major developments and even minor changes to local conditions on planet earth, viewed with such foreboding by the public and reported with promises of imminent doom by the media?

Where is the spirit of human optimism and the belief that changes made, even in a delicate and sensitive ecosystem, can bring real, positive improvements in quality of life

1

and the environment? Furthermore, surely the potentially negative view that wastes and pollutants resulting from human activities are intractable, can be replaced by innovative technologies that break down, recycle and successfully manage these same products?

Thanks to agitation by informed members of the community, legislation in the European Economic Community (EEC), US and Australia now requires all developers not only to submit an Environmental Impact Statement (EIS), but to allow adequate time for public comment. Unfortunately, few of us know enough about an EIS to be able to comment - even if we are not happy about a proposal. Any resultant 'comment' tends to become a hotch-potch of irrational and ill-informed emotional statements lacking substance and organisation.

At other times, 'concerned' individuals enlist the assistance of committed environmentalists, who are involved in broader issues and do not assess the true merits and demerits of the proposal in a balanced manner, seeking only to de-rail the project for political gains on other fronts.

In this book, therefore, we aim to empower you, an average member of the public, with the means to understand an EIS and, through your understanding, to equip you with the tools to make worthwhile, constructive and substantive comment on any development proposal which may affect you or those for whom you feel responsible.

The History of Environmental Impact Asessment - EIA

Environmental assessment, as it is known today, first began in the United States around the time of the Second World War. This period corresponded with the large-scale growth of the pharmaceutical manufacturers, intensive agriculture using toxic chemicals, major engineering and mining works with the potential to radically alter the natural landscape, and, of course, the threats and benefits represented by nuclear power.

Introduction

Previously, the products and wastes produced by human industries were not judged to carry a high toxic risk. Nor were they used in such large quantities or considered likely to remain, in their active form, in the environment. The atmosphere, rivers, soils and oceans had proved capable of absorbing and recycling human wastes without irreversible damage.

Until the Industrial Revolution in Europe, humans had neither witnessed nor been able to perpetrate, on the same scale, any major long-term damage to their environment which nature could not, in time, mitigate and reverse. This includes localised and even extensive clearance of timbers and land for agriculture such as occurred in Central Europe and Northern Australia. Large forested areas still remained. Food scarcities, disease and vagaries in climate placed natural constraints on human survival and maintained a delicate balance between resources and their exploitation by human habitants.

It was not so much the Industrial Revolution itself as the population boom it spawned and the associated social changes, which initiated the present problems. Exponential growth in human numbers, greater exploitation of resources by individuals and new powers over nature, have far outpaced the slower rate of natural adjustment and normal evolutionary change.

The whole process is now out of synchrony. The result: vast mountains of toxic wastes which fail to break down and recycle by natural means; the emergence of new viruses and micro-organisms which defy the human immune system; the imminent extinction of biological species, and the depletion of a long list of already dwindling resources such as pure water, rainforest hardwood and trace minerals.

Yet too often the blame is placed on industrial development *per se* without adequate recognition that it is the human beings themselves, their numbers and their demands, which have created this situation. It is they who have empowered the industrialists and established the reward structure which links development, profit and power.

All ecosystems - including human beings - have thresholds of tolerance for their own numbers, their wastes or pollutants and the disturbance they cause. Beyond these thresholds, the system may suffer anything from temporary upsets to complete destruction.

Table 1: Underlying philosophy of environmental impact assessment (EIA)

Decision makers have a responsibility to involve and support the participation of the public in decisions which affect the environment.

There is a link between the health and wellbeing of all living species, human, animal and micro-organic, with the environment.

Physical, social and ecological aspects of the environment are all relevant to human health and wellbeing.

Environmental impact assessment is an implicit and essential element of ensuring ecologically sustainable development of all resources within the environment.

Table 1 summarises the relationships between data compartments which constitute the environmental impact statement (EIS) and its subsequent assessment or review.

The Development of Procedures for EIA

Public concern about the environment lagged behind the worried noises of those few concerned scientists not swept up in technology worship. Rachel Carson's sobering book *Silent Spring* was published first in USA (1962) then in UK (1963). She succeeded in making the public aware of the ecological consequences of introducing toxic chemicals into natural food chains, and the dire effects of cumulative dosage with apparently small quantities of agricultural poisons.

By the end of the 1960s, pressure from the public and environmentalists had forced state and federal authorities in USA to impose some controls over the release of toxic chemicals into the environment. The control was established by the 1969 National Environment Policy Act, which required Environmental Statements to be prepared for federally funded or supported projects which were likely to have impacts on the environment. The US Council for Environmental Quality was charged with the task of developing standard procedures for Environmental Statements. While these controls were substantially eroded under the twelve years of the Reagan and Bush administrations, the 1993 Democrat Presidency has a mandate to revive and strengthen environmental controls.

A Global EIA Process

Environmental assessments vary enormously in size and scale. The sensitivity of the area, the degree of disruption likely to be caused and the project itself are all relevant. There is increasing public pressure for the introduction of intercontinental and global Environmental Assessments. The increasing level of environmental hazards with global implications, range from ocean dumping of toxic and nuclear wastes, to acid rains, over-fishing, desertification, and the implications of rainforest logging with greenhouse effects, ozone depletion and losses in biodiversity.

International attention is also drawn to environmental risks and damage on large-scale projects such as World Bank funded schemes like the Aswan Dam (Nile River, Egypt), and the Sardar Sarovar Dam (Narmada River, India), plus controversial schemes such as hydroelectricity associated diversions of the Danube River in Eastern Europe. These major projects affect large numbers of people and cross national boundaries. A uniform set of international legislation and procedures is warranted to protect the remaining resources from human greed and exploitation.

The European System of Environmental Control

The European approach to environmental control owes much to the United States, where they developed the first viable procedures and pointed to the weaknesses inherent in any system. For too long they relied on litigation over pollution claims subsequent to completion of a major chemical or processing plant. This necessitates a time-consuming accumulation of scientific and technical evidence to substantiate the claim.

The EEC has now established the concept and practice of carrying out environmental assessments for major projects before they are constructed, rather than relying on international law and public outcry to limit and control environmental damage. Pollution and its consequences are emphasised with acknowledgement of the impact of projects on employment, social structure and economics, all issues likely to be associated with local rather than international concerns.

The EEC Directive gives the main priorities for assessment as human health, the quality of life as it is affected by the environment, the continuing diversity of species, and the maintenance of the whole ecosystem. Its focus is on human beings: the goal is continued but "managed" human prosperity. Absent is a recognition of the dependence and centrality of plants and animals to human survival and wellbeing. Attention is needed to the implication of projects and developments on species inter-relationships which maintain the soil, water and atmospheric cycles in the natural environment. The atmosphere, the hydrologic cycle, global weather and climate receive but passing attention with a focus only on microclimate (local effects). Changes to the genetic heritance of species, and altered survival patterns associated with loss of habitat, all need to be considered.

Introduction

Measures for remediation, management of damage and restoration of contaminated sites all need strengthening.

There are other matters, too, which require resolution by member nations. Legislation on free trade prevents any member nation from denying passage of any other member's goods across its borders. This serves to facilitate rather than restrict transport of toxic wastes, and encourages dumping in poorer countries where information on government policies and practice is not readily available. River networks such as the Rhine and the Danube, which cross a number of natural boundaries on their way to the sea, carry with them products acquired upstream to countries and people living downstream. Thus the wastes gained upstream pollute the water and affect wildlife survival in waters further downstream.

The Public Interest

Where the public is aware of possible impacts, the level of opposition to a project depends on the current state of public opinion rather than on a long-term and unbiased appreciation of the environment. In an economy where there is a significant level of unemployment, the effect of a project is usually to increase the number of jobs available and to generate spin-offs in the form of service and supportive industries. There is unlikely to be any major opposition from the local population or trade unions, and protesters will be roundly condemned and accused as idealists or 'stirrers'.

In Europe, the impacts which usually attract opposition are those on the visual quality of the landscape, the pollution and disturbance of the ecology of the area, the land-take of houses and agricultural land, and the effects of new infrastructure on human activities. In Australia, forest logging projects for the export woodchip and paper industries, and new mining ventures likely to threaten a fragile arid environment or traverse local aboriginal 'sacred sites', are frequently the focus of major environmental impact assessments.

Legislation in principal westernised nations today is designed to ensure that the public has an opportunity to comment either during the preparation of an EIS or before it is accepted by the responsible authority. Opportunities can be maximised to ensure that measures to mitigate any undesirable aspects of the project are included if community concerns can be addressed early. At the same time, there can be included compensatory clauses and provisions for minimum discomfort, disharmony and delay during the construction phase.

Unfortunately, the people in an area likely to be affected by a proposed development often learn of it first from gossip and rumours. Alternatively, they may read an emotive article in a local paper focusing on the concerns of one affected individual who may not always possess the true facts. First impressions tend to create doubts which are then fuelled by negative feelings and a general distrust of 'developers' as a whole. Resistance to change introduces the final element to what ultimately materialises as public opposition. It is therefore essential that the public in general, and the local community in particular, is fully and accurately informed of a proposal at an early date. Information should be readily available, detailed but understandable. Thus, at all stages in the development process, an intelligent and balanced approach rather than one clouded by emotionalism, distortion and conflict should best serve the interests of all concerned and most benefit the community as a whole.

Concluding Remarks

It should be realised that the environment is not a static situation, and that many changes have always occurred and will continue to occur without the intervention of human activity. Many of the projects opposed by so-called environmentalists have wrought changes but have also promised real gains to the quality of the environment and its resources. There have always been species which have become extinct while new ones evolved.

It is unfortunate that so many protests are raised before the full facts are known and have been studied by all parties. Losses must always be balanced against gains. There will always be winners and losers. It should be our role to ensure that the good guys are helped to be winners: health and happiness for the majority means moral issues are paramount.

Thus, this book attempts to improve the understanding of the process of sustainable development through knowledge about the goals, nature and formulation of the environmental impact statement and its review processes (the assessment phase). It should be used also as a resource tool and reference for those wishing to take advantage of the opportunities that now exist for individual members of the public to significantly affect the environment. Knowing what an environmental impact statement says, and how it is structured, makes it possible for any individual to express an opinion about a development project in a form that demands consideration and resolution.

CHAPTER 1:
ENVIRONMENTAL IMPACT ASSESSMENT
What is it? Why do it?

Environmental impact assessment (EIA) can be described as a process identifying the consequences for the total environment of undertaking new developments and changing natural systems. It arose to resolve the competing demands for economic development with preservation of the quality of all life in the surrounding areas.

Initially, the EIA process aims to provide decision makers with scientifically researched and documented evidence to support a reliable prediction on the likely consequences of their proposed actions. The outcome is a formal document or report referred to usually as an environment impact study, statement (EIS) or appraisal.

The term 'impact' and 'effect' are frequently used synonymously. An 'impact' describes changes in an environmental parameter over a specified period and within a defined area resulting from a particular activity. These changes are compared with the previous, baseline situation which applied before the activity had been initiated. An impact varies over the area and with time; both should be considered.

The site is described by data prior to the commencement of the project, and any subsequent changes are monitored against that data. Hence, an 'effect' represents a specific measurable change in the site.

Consultants differentiate between the effect of nature and the effect of human activity on an area and assess their respective impacts. Predictions of the impacts depend on assumptions made about the rate and direction of natural changes which are used to construct models resembling the conditions and systems under study. They necessarily contain limitations. Decision matrices, network approaches and statistical analyses are all used in an attempt to quantify relative risks and probable benefits associated with a proposed development.

Basic Features of an EIS

Environmental impact studies prepared for individual projects generally embrace the following considerations:
1. description of the main characteristics of the project
2. estimation of the residues and wastes that it is likely to create
3. analysis of the aspects of the environment likely to be significantly affected by the project
4. analysis of the likely significant effects of the proposed project on the environment
5. description of the measures envisaged to reduce the harmful effects
6. alternatives to the proposed project together with reasons for rejection
7. assessment of the compatibility of the project with environmental regulations and land-use plans
8. systems for monitoring project and post-project
9. non-technical summary of the total assessment
10. recommendations (if sought)

The majority of EISs assign or rank impacts in terms

of the severity or disruption likely, should the proposal go ahead. Primary (major), secondary and tertiary (indirect) impacts can be distinguished. Considerations of time and the rate of natural changes, mitigating effects and the unpredictable elements of behaviour of natural systems following a development are all important. For example, it might be expected that forest logging and increased human traffic into a formerly pristine and naturally wooded area might threaten survival of native species. Also conservationists engaged in protests against hydroelectric power developments in NorthWest Tasmania (Australia), land clearance in Sarawak (Malaysia) and rainforest timber logging in Daintree (Northern Australia) frequently identify local species whose environment will be altered. Their hypotheses which may promise imminent extinction of a species, need to be backed with scientific research. General arguments based on unproven claims imply that any change to a natural environment destroys natural nesting, feeding sites, and species interdependencies which have evolved over thousands of years. This is not always the case.

It is equally true that recognised changes in one environmental niche do not necessarily reflect the only likely course of events to be anticipated by similar changes to an apparently comparable environment elsewhere in the globe. Each niche in itself is unique: each proposal, its geophysical location and its affected population, human and non-human is unique and needs to be considered individually and independently.

In the general sense, changed conditions in a pristine environment facilitate the entry of new species which, in the absence of natural predators can compete against a native species for scarce environmental niches. New diseases, new predators, and a loss of biodiversity result. But are these effects all to the detriment of planet earth? In the long evolutionary history of most species, co-existing organisms such as human beings and their activities would exert selective pressure which would contribute to the evolution of new improved species. These newer forms of life may be better adapted to survive in the changed environment than earlier, older forms which they eventually displace.

The reality also is the recognition that there are few, if any, places remaining on planet earth where human beings have not left their footsteps or their mark. Even the ice atop Mt.Everest has been shown to contain dirt traced to factory emission from northern Europe.

Procedural Elements in the EIA Process

The main procedural elements of the EIA process will normally involve the following steps:
* the developer calls together consultants, regulatory bodies and other organisations;
* independent panels vet the studies made for major projects;
* the study is published and is used as a basis for consultation involving both statutory authorities possessing relevant environmental responsibilities and the general public;
* the findings of the consultation process are presented to the competent authority reviewing the project;
* any mitigating measures and claims for compensation are considered;
* progressive and post project monitoring of environmental consequences arising from implementation of the project are set in place.

The developer and his team prepare an environmental impact study which is submitted along with the project application for project authorization, to the competent authority. Checks and balances exist to monitor the consultant's activities and assess the quality and integrity of data contained within the resultant report. Members of the independent panel involved in the evaluation include academics and professional scientists, overseas experts in the field and parties who have been associated with similar projects in the past.

A clear distinction should be drawn between techniques for predicting individual changes and those appropriate to EIA. For example, ground level concentrations of atmospheric pollutants can be calculated using Gaussian dispersion models, but more comprehensive and complex predictive models are required for assessing environmental impacts over a whole site.

It is also important to remember that each site and proposal is unique in itself and flexibility must be maintained to deal with new, different issues as they arise.

By way of illustration: the effect of a proposal for a small gold mine in a country town with little employment may seem an excellent source of welcome employment for its inhabitants. However, if the mine is located in Australia, it may infringe on an aboriginal 'sacred site', a traditional hunting ground or wandering of a dreamtime ancestor. Any proposal to develop the area is likely to be viewed with animosity and resentment by the same population who might be presumed to benefit in an occupational and financial sense. Experience in anticipating such future constraints and impacts upon the proposed development would not figure in EIS guidelines deemed acceptable in, for example, Germany or England.

There is a further consideration which is often not well accommodated by the EIS process. Environmental systems adjust and change with the conditions and under human influence. Some are very dynamic, others appear more or less static. This means that, regardless of dire predictions, assumptions made may prove unfounded because of unanticipated natural adjustments once the project is operational and project work has been completed. The ability of a special variety of pine to germinate, survive and flourish naturally on denuded and highly contaminated mine tailings in abandoned coal and mineral tailings dumps in eastern Europe, has confounded British scientists who have failed to achieve successful restoration of the landscape in coal mining regions of Wales and the Midlands using traditional land reclamation techniques.

Legislation

Legislation specifying and defining the need for EIA varies between states within countries and from nation to nation. It lacks uniformity in international terms and fails to apply to many areas where environmental impacts and effects are most evident and damaging in a global sense; e.g. contribution to ozone depletion and greenhouse gases caused by large forest burnings in Amazonian Basin.

The present framework for EIA legislation resulted from the town planning and zoning movements of the early twentieth century, and accompanied rapid and uncontrolled urbanisation in the United States, England and Germany (Woodhead, 1990). Individual countries and separate states within them have legislative and planning powers which overlap, duplicate and counter the laws which have been agreed to internationally, by federal and state legislative bodies and government planning authorities. Disputes about jurisdiction hamper action, and an adversarial climate is created where conflict is continual and unabated.

For example, in the United States, the Council on Environmental Quality (CEQ), the Environmental Protection Authority (EPA) and numerous individual state authorities are all active at the federal level in appraising major proposals. State instrumentalities apply their own rules which are not always consistent with federal laws. In the United Kingdom, legislation on environmental impact comes under the control of the European Community's Environmental Impact Assessment Directive rather than the Parliament in Westminster.

Australian legislation was overhauled after the 1960s, but twenty years further on the EIA process still contains areas of anomaly and contentions. For example, it is only in some states that planning and development rules require all projects over a minimum size and cost, to be reviewed and evaluated by an independent, external consultant. Separate pieces of legislation remain which specify different procedures and have a different scope or focus. Current initiatives of the Australian and New Zealand Environment and Conservation Council's (ANZECC) seek to develop a uniform approach to EIA to apply to both countries and based on international proposals.

Problems which remain for legislators on the environment include:

(i) inconsistencies between individual regions or states within national borders; e.g. Wales, England and Scotland

(ii) intervention in EIA by other governmental bodies

(iii) inadequate mechanisms for enforcing environmental regulations; i.e. who polices and enforces regulations?

(iv) inadequate integration between environmental planning and continuing environmental management and pollution control

(v) poor mechanisms for incorporating science into EIS

(vi) lack of accessible repositories of information on previous impact predictions

(vii) limited formal involvement by the public

(viii) inadequate auditing and testing of the predicted impacts

(ix) inadequate frameworks for integrating the economic and non-economic considerations in development plans

(x) inadequacy of existing procedures for resolving conflicts: means for establishing alternative and harmonious resolution of disputes

(xi) inadequate attention to waste management and recycling options within EIS.

Health as an Issue in EIA

It is generally conceded that health is a specific but implicit concern in environmental impact assessment. Many public health professionals and public interest groups believe that a special health impact requirement should be incorporated into the formal assessment procedures legislation. Their rationale is based on the fact that unless specific attention is directed to all concerns related to health, only the major obvious issues will gain attention.

For example, in a proposal for a new international airport or a major highway, the effects of increased noise levels on hearing and sleep patterns of residents would be the focus of research data and tests. Associated health issues including toxic atmospheric pollutants like nitrous and sulphurous oxides, lead and carbon byproducts must receive mention. A full health impact survey would identify and research thoroughly all the potential risks and benefits to health associated with a proposed development.

There are other reasons which enforce demands for a separate and specific section in an EIS, dealing with health. These lie in the present manner in which health issues tend to be incorporated into the assessment of environmentally sensitive projects.

For example, Australian legislation fails to cover issues such as health risks to swimmers as a result of water pollution arising from deep ocean outfall for primary treated sewage located three to six kilometres from Sydney beaches (Sharon Beder: *Toxic Fish and Sewer Surfing*, 1989). Health issues associated with the carriage of large cargoes of toxic nuclear and non-nuclear wastes on international waterways, into ports for servicing of the vessel and through waters known to be hazardous, is not incorporated into existing national or international agreements. Disposal of hazardous chemicals under commercial contracts negotiated between a source in one country and a servicing agency in a second, recipient country can also involve passage through areas where individuals are at risk.

Community concerns created by media speculation, not always based on scientific evidence, may also demand comment on health effects. Air pollution levels accompanying increased industry and motor vehicle transport use in the area are frequently and incorrectly blamed for increased hospital admissions due to asthma, when studies can confirm only a weak link (*NSW Public Health Bulletin*, August 1991, 2,8:72). Weather conditions and seasonal factors may be the true causal agents.

Final Remarks

All these tend to generate high levels of emotion associated with real or perceived risks to health and well-being occasioned by even the most benevolent of developments. There is a common tendency for the public, assisted by the media, to view any projects advanced by business, and, any 'radical' venture as unacceptable. Consequently it is rejected more or less by reflex. As a corollary, all efforts to resist developments, whether soundly based and logical or based on emotional arguments alone, as evidence of public concern are highly admirable, and ethical actions are viewed as 'good'. Such efforts are praised by virtue of the fact that they are seen to represent the attempt of the 'little man', the powerless to resist subordination of his interests to the 'big man', the powerful.

Unfortunately this means that the environment becomes the stage on which individuals play out their own personal sense of power-lessness in the face of high levels of unemployment, technological change, social and family breakdown. It is important to remember that substantial and positive improvement often results from projects and must always be balanced against adverse features. Irrigation projects, drainage works, flood mitigation and even the petrol-guzzling motor car all bring benefits as well as negative effects.

They represent change: perhaps it is change which people fear most, and it is change *per se* which they resist.

CHAPTER 2:
'BACK TO BASICS':
The components to an EIA

333
1

Introduction

Chapter 2 is about 'basics', the key components to environmental impact assessment. In simple language, it aims to introduce the four main categories of data comprising an EIS - physical, eco-biological, social and health information. Technical data falls into these four categories.

When an environmental impact assessment is undertaken, the project proponent engages technical professionals who collect data on the nature of the site, the project and the local environment. These professionals organize and rank this data, in importance according to the type of development involved, and summarise their findings in an environmental impact statement (EIS). This EIS is incorporated into the project application which is then placed before planning authorities for approval.

By way of example, an environmental impact study for a large dam for hydroelectric power generation and irrigation would report extensively on flora and fauna in the area to be covered by water collected behind the dam wall. On the one hand, re-settlement of people living in the pro-

A002657

posed catchment may be required; vulnerable indigenous plant species may be threatened by alteration of a unique and fragile habitat. At the same time, new employment opportunities will accompany the establishment of a power station; improved road access would bring improvements in the quality of life to residents; the planned dam could serve as a protected breeding ground for threatened native fish and wildlife species.

A proposal to build a paint factory in a semi-urban area would identify other factors of importance and relevance to the existing and future environment. Because workers would be needed for its operation, the factory would be planned near people. It would need to be located nearer transport corridors to facilitate cheaper, ready distribution of paints for sales and export markets. The environmental impact statement (the EIS) would focus more on people-related issues such as emissions, economics, wastes, social factors and health before effects on native animals and plants.

In other words, the extent to which physical, social or biological data are central to the EIA will depend fundamentally on the project or development under consideration at any one time. This becomes apparent as each assessment team gains experience of any site, and that is added to by more data collected from comparable sites and assessment processes.

Knowing What Is Necessary

There is a uniform format under which an EIS is prepared and this was discussed in the previous Chapter. Nonetheless, the professional, or simply any interested party such as a group of neighbours living in an area where a large sports complex is proposed for a bushland reserve containing native wildlife, will need to ensure that their own study or the proposal they are querying conforms with the requirements of the law in their country, state or local area.

Back to Basics: The Components of an EIS

The regulations applying to the production of an EIS in the United Kingdom and member EEC countries contains three Schedules. Legislation in the United States and Australia is similarly drafted.

Schedule 1 lists projects such as large-scale engineering works for which EIA is compulsory (see Table 2.1). Schedule 2 lists other projects including extractive industries and developments which may have a significant environmental effect (Table 2.2). These Tables are quite detailed and comprehensive in the range of industries and projects specified, but the planning authorities still possess power to specify and demand inclusion of an environmental impact statement where they deem it necessary.

Table 2.1: Projects subject to mandatory EIA
(after Waldern, 1990, p.10)

Extractive Industry
extraction and briquetting of solid fuel
extraction of bituminous shale
extraction of ores containing fissionable and fertile material
extraction and preparation of metalliferous ores

Energy Industry
coke ovens
petroleum refining
production and processing of fissionable material
generating of electricity from nuclear power
coal gasification plants
disposal facilities for radioactive wastes

Production and Preliminary Processing of Metals
iron and steel industry, excluding integrated coke ovens
cold rolling of steel
production and primary processing of non-ferrous
metals and ferro-alloys

Environmental Impact Assessment

Mining and Oil Exploration
mining and preliminary processing of metal ores by open
cut and underground techniques
extractive activities conducted on mining sites
coal and petroleum exploration on a large scale

Metal Manufacture
foundries
forging
treatment and coating of metals
manufacture of aeroplane and helicopter engines

Manufacturing of Non-Metallic Mineral Products
manufacture of cement
manufacture of asbestos-cement products
manufacture of blue asbestos
manufacture of products derived from sand minerals

Food Industry
slaughter houses
manufacture and refining of sugar
manufacture of starch and starch products

Processing of Rubber
factories for the primary production of rubber
manufacture of rubber tyres

Building and Civil Engineering
construction of motorways
intercity railways including high-speed tracks
airports
commercial harbours
construction of waterways for inland navigation
installation of pipes for long-distance transport

Chemical Industry
petrochemical complexes for the production of olefins,
olefin derivatives, bulk monomers and polymers
chemical complexes for production of organic intermediates
chemical complexes for production of basic inorganic chemicals

Back to Basics: The Components of an EIS

Table 2.2: Projects to be examined to assess the need for EIA

All new developments in existing communities, especially where significant human and animal populations exist or where vulnerable organisms, physical systems or communities are likely to be found.

Opening of new settlements or areas which have the potential to damage ecology, reduce air or water quality:
e.g. siting of a small sugar-milling factory plant on margins of sugar cane growing area, establishment of any permanent human settlement in fragile environments.

Proposals and developments which threaten fragile or vulnerable populations or undermine biodiversity or organic species within the area: e.g. decision to partially remove poor quality timber species from forest.

Proposals to decommission or re-zone lands formerly used for industrial or agricultural purposes, and especially with a view to future human settlement and urban space use: e.g. former site for cattle tick dip, former gas or chemical storage facility, former mining site, etc.

Developments which increase human activities and presence in the area: e.g. fire access trail into forest or conservation area, tourist expeditions.

Developments which generate noise or increased traffic flow: e.g. upgrading of road from local suburban to general usage, siting of playing fields in waste lands.

Developments which generate a high level of community concern: e.g. transmission lines, freeways, airports.

Developments which have the potential to expose workers to hazardous products and wastes in the workplace: e.g. paint manufacture, cotton dyeing, printing, bleaching and other chemical, nuclear or petroleum industry, mining, etc.
Large-scale traffic, transport and communications movements

which could pose a threat to local environment and populations: e.g. movement of oil tankers through hazardous marine waterways or into areas where unique, fragile biological ecosystems are found: e.g. Barrier Reef.

Existing or proposed commercial ventures of a local or international nature which could threaten local or global food stocks and scarce resources: e.g. large-scale harvesting of rainforest timbers, drift net fishing, whale harvesting and fishing.

Schedule 3 sets out the type and scope of the information which must be included in any EIS. This comprises data which will enable the local planning authority to identify and judge the environmental effects of the project. It specifies the need for:

* a physical description of the site
* environmental data that would permit the planning authority responsible to assess potential environmental effects of the project
* the provision of technical data in the following categories:
>   (a) the ecosystem
>   (b) climate
>   (c) water
>   (d) soil and the land
>   (e) flora
>   (f) fauna
>   (g) human beings
>   (h) areas of cultural significance or heritage

Each environmental sector should be described or assessed and effects on it discussed.
* details concerning methods which can be incorporated to minimise or modify significant adverse effects, including measures such as screen planting of vegetation to improve the visual landscape and filtering systems to limit pollutant emissions.

The remainder of this chapter concentrates on Schedule 3 and the techniques for describing and collecting the required data.

Back to Basics: The Components of an EIS

Physical Description of the Site

The first task for the consultant responsible for preparing the EIS is the collection of baseline data about the existing site and its inhabitants, plant life, animal and human beings.

Step 1. The physical description of the development site.

This involves:
* a physical survey of the existing site
* description of existing land uses
* existing developments on the site, their structure and uses - e.g. houses (residential), churches (religious observance), reserves (wildlife conservation)
* special designations affecting the site or its neighbourhood, including ancient monuments, areas of outstanding natural beauty, areas of archaeological importance, conservation areas, heritage areas, environmentally sensitive areas, National Parks, green belts, National Trust or Estate listed buildings, traditional or sacred sites, etc.
* contours and existing man-made and natural features
* existing infrastructure: roads, rail, sewerage, water, gas, electricity

Step 2. The proposed development and its effects on the physical nature of the site.

This can be broken up into three main data sets - pure physical, production processes and estimated emissions and residues from the development:

(i) pure physical
* land-take for the construction period
* land-take for permanent development
* land-take for reserve for future development
* land-take for ancillary development, housing, recreation

and community facilities
* land-take for infrastructure resulting from development
  - e.g. new roads, water reservoirs, electricity sub-sta-
  tions, hospitals, schools
* land-take for amenity: screen planting, screen moulding
* time-scale of development

*Table 2.3: Land development projects - probable impact
categories*

Land Economy
public fiscal balance
employment
wealth

Aesthetics and Cultural Values
attractiveness
view opportunities
landmarks

Natural Environment
air quality
water quality
noise
wildlife and vegetation
natural disasters

Social Impacts
people displacement
special hazards
sociability/friendliness
privacy
ethnic/racial groupings

Public and Private Services
housing
water supplies
sewage
police and crime
education
public transport
motor vehicle movements

municipal councils - wastes
energy services
shopping centres
hospital services and care
fire protection
recreational facilities

   Physical impacts which normally attract opposition are
the impact on the visual quality of a landscape, the pollution and disturbance of the ecology of the area, the land-
take of houses and agricultural land, and the effect of new
infrastructure on human activities.

Table 2.4 Data set for urban development proposal
* composition of the community
* community organisation, structure and resources
* community capacity - commitment, mobilisation
* physical environment - housing, transport, open space
* socio-economic environment - incomes, employment,
    education, literacy, types of housing
* disease and disability profile - major health problems
* health and environmental services - service use and
    acceptability
* social services - service use and acceptability

(ii) production processes
* if industrial, capacity of plant, processes used, raw
    materials required
* energy requirements: gas, electricity
* natural resource requirements: water, timber, quarry
    materials, land-fill

(iii) estimated emissions and residues from the development
* notifiable pollutants
* emissions to air
* wastes discharged to water, soil, or disposal sites
* wastes for special disposal (toxicity)
* wastes for potential recycling
* noise levels during construction (day and night)
* noise levels during operation (day and night)
* heat emissions (day and night)
* vibration during construction (seismic testing before
    construction should be included)
* vibration during operation
* radiation levels
* light levels (night operation)

Step 3. The after-use of the site.

This mainly concerns operations involving mining and instances where sites are re-zoned following industrial use:
* removal of plant and buildings
* proposed land uses after clearance
* toxic residues assessment and review of site
* detoxification measures (if warranted)
* time-scale for restoration

The Ecosystem

Directly or indirectly, every part of the natural world is dependent on every other part. The survival and continued function of the system itself rests on maintaining the balance between production, consumption and re-cycling of the integral and inter-dependent parts.

These parts are generally referred to as living and non-living. They are defined more specifically by their properties, and these properties serve as useful tools for organising data. Of the non-living components, a gas such as nitrogen and a liquid such as water, for example, are physical components of the ecosystem. Rocks, mountains and soils are geological structures. Pesticides, petrol and sulphur dioxide emissions are chemicals.

Yet despite the effort at simplification, there are areas of overlap. Lead is a physical element: it is a metal and it is a natural component of the earth's surface where it occurs in geological structures. It reacts with another natural and physical component of the earth's atmosphere to form lead oxides, and in this form is present all over earth's surface.

The physical, non-living components of the system themselves interact to produce climate and weather. They create the environment which promotes, facilitates or discourages the emergence and perpetuation of living forms. Thus apparently inert parts of the system exist in subtle and balanced relationships with the other part of the system: living beings.

Flora (plants), fauna (animals) and human beings comprise the three main categories used to describe life-forms on earth. Micro-organisms such as fungi, bacteria and viruses are classed as separate entities, although they are sometimes labelled as members of the plant or animal kingdom. All these organisms interact with one another. Indeed, their existence is dependent on a web of mutual relationships which ensure survival of the fittest members in the total environment at any one time, and as determined by the physical environment or conditions in which they find themselves. The whole comprises a neat, highly integrated and closely interdependent series of relationships in which the survival of any one is crucially dependent on its neighbour, and is influenced by changes in any one or more of the environmental conditions.

The Climate

There is a close relationship between the atmosphere and the physical, chemical and biological components of our earth. Change, growth, development and perhaps even the initiation of life itself require the presence of basic materials, a favourable environment and a supply of energy. The climate is of fundamental importance. We must define and study it in EIS to evaluate any changes which may occur by human or outside actions.

Species develop in response to climate, and the more robust types adapt to their surroundings and survive its many changes. Other varieties migrate, develop different characteristics, multiply or die as the environment changes in time and space. There have always been contests for the more favourable areas, with winners dominating the scene and losers moving and adapting to the less congenial regions, or disappearing.

Human beings have been the most successful of all animals in adapting to conditions on the planet and modifying it to meet his needs. Together with plants and animals, they wage a constant battle against the weather, their principal adversary. To survive they use a favourable climate or seek protection from its extremes. Their many migrations and travels show that the human species can live in a wide range of climates, and can populate most land areas of the globe.

Weather on this earth is a consequence of the immense forces of solar radiation and celestial movement. The earth's weather is affected by the sun, the earth's shape and rotation, the atmosphere and landforms on its surface. Planet earth and its thin envelope of air rotates as it moves in an orbit about the sun, its prime source of energy. Each movement within the stream of radiation from the sun, and the shelter of the various components of its air envelope, result in changes of input and output in solar energy. This in turn gives changes in the weather.

The earth intercepts an infinitesimal proportion of the sun's output of energy. It follows that, in human time, there are no large variations in the receipt of solar energy. The standard amount is a solar constant. With solar flares and sunspots the quality of the radiation changes as witnessed by radio interference and auroras. The atmosphere with its contained gases filters and reflects back part of the solar radiation. The remainder warms the earth and its atmosphere, which in turn transmits back long-wave radiation.

Temperature changes on earth give rise to a global circulation which can be referred to as the macroclimate; in contrast with the microclimate, the climate of the immediate surrounding soil, plants and objects. Microclimate is adversely affected by increased industrial emissions and high-rise buildings, at the same time as it is improved by increased vegetation through shelterbelts and plantings over the site area.

In the period of a human lifespan, the macroclimate

is intractable and displays few variations. Long-term changes have coincided with Ice Ages and periods of warming. These are probably the result of variations in the earth's atmosphere precipitated by volcanic activity, dust, gases and moisture change: even the effect of impacts from asteroids and other stellar debris.

Probably the component of the ecosystem which people notice first is a change in immediate living conditions: the presence or absence of trees, changes in the physical landscape, hotter, drier weather, increased amounts of smog, the presence of fumes, etc. Most people accept that it is difficult to change the macroclimate. The smaller, more immediate changes they observe occur within the microclimatic zone. These arise both naturally or as a result of human intervention. They can improve as much as degrade the environment.

For example, a microclimate can be established within sheltered areas by control over radiation, wind, temperature and humidity. This can lead to physical, chemical and biological activity which accelerates by interaction until a wide diversity of conditions are set up. Living creatures, as part of the picture, give their input and influence, each contributing their part. Small islands of inter-dependent living species can begin with attempts based on creating a local microclimate. These small ventures can initiate large-scale regeneration of a larger area. They are of value in reclamation of damaged and degraded lands, particularly those ravaged by drought, over-grazing or excessive removal of vegetation and tree cover. Other measures can be taken to minimise adverse effects and to further improve the environment.

Some simple examples of ways this may be achieved are:
* setting up of shelter belts of trees of varying height and canopy density
* siting of buildings and other structures

* reafforestation and revegetation of large barren areas at risk of dust storm, tornadoes and wind eddies
* restricting radiation input by the use of meshes, shading, orientation of crops, close planting and inter-cropping
* controlling water loss by measures to limit evaporation and radiation loss
* selective irrigation, such as better and more economical use of drip and sub-surface water supply
* control of species numbers and demands: e.g. through removal of introduced and feral members, directly responsible for or indirectly contributing to higher water demands than would be the case in the natural, pristine environment.

The above examples suggest that although humans cannot exert control over the broad scale circulation and properties, they can modify climate near surfaces or within the microclimate. They provide shelters and enclosures, introduce water into the system, provide and release energy and materials from other resources. In fact over small spaces almost any type of climate or conditions can be created.

It is not only within the atmosphere that changes can be made but also within the hydrosphere or within the earth itself. It follows that a very careful examination and documentation can be made of the existing microclimate within a region and from then, its potential for change.

Thus it is important to consider changes in the micro-climatology by the human race and their implications for all species. These changes may not always be deliberate. Nor should it be assumed that they are always adverse and contribute to degradation of the environment or loss of species. The quality of the environment can be immensely improved for plants and animals as much as for humans, in particular by attention to areas where natural processes may be responsible for significant and on-going despoilation. The building of levy-banks and the construction of culverts and canals can serve to limit physical damage to the landscape, loss of topsoil, animal and plant life which accompanies annual flooding of monsoonal river deltas.

At the same time, harmful effects may occur long after the human activity has been in operation, and only after some difficulty in identifying the actual cause of a problem which has been detected in one or other species community. Human cancers and genetic changes linked to the presence of foreign chemicals are but one example which is evidenced only over time and with the birth of a new generation. It is because of these unexpected and untoward occurrences that it has been seen imperative to introduce a constraining, evaluating and potentially predictive component to the planning process in the form now called Environmental Impact Assessment.

The meteorologist or atmospheric physicist on the impact assessment team focuses on both macro and microclimate. In the first instance, local measurements of radiation, temperature, humidity, precipitation (rainfall), wind force and directions are sought from the nearest available monitoring station. A basic description of the climate at various times of the year is developed. Further readings and observations are taken at various points over the site area and serve to further focus the historical data.

For any EIS this data is essential to give basic conditions. Averages, frequency distributions and extremes over some period of time are sought. Statistical analyses and models play a part in defining trends and predicting scenarios.

With this basic data it is possible to further introduce changes which would be expected to accompany the project under consideration. This means that the meteorologist needs to work in close collaboration with all other members of the team and be aware, in particular, of changes in the topography associated with construction, increases and decreases in gases and wastes as a result of industrial activities and effects occasioned by civil engineering and construction: e.g. roads, housing, etc.

Developments which warrant attention include those undertaken in areas where climate is unstable, or topography is likely to cause pooling of smog and toxic emissions from industry or traffic. Others may generate high thermal outputs, such as large industrial complexes and power houses. Increased levels of urbanisation, high population density and a significant increase in high-rise building can result in unusual wind tunnels, heat islands and other impacts likely to affect human health.

People already at risk from heart and lung conditions are most affected by increased levels of smog and sudden changes in temperature. Others experience psychological problems working in high-rise, air-conditioned buildings. Infectious organisms including pneumococcal, streptococcal and influenzal species flourish in damp, moist environments such as air-conditioning plants and cooling towers common in urban high-rise buildings.

Water

Developments tend to impact on the water sector of the environment in both a physical and a chemical sense. A comprehensive description of the water sectors should identify all permanent and temporary sources, including those present on a seasonal basis only. In this way and many others, there is close interaction between the water sectors, climate and living species.

Reference to water sectors in an EIS should refer to all instances of static and running water (e.g. rivers, streams, ditches), including those areas devoted to recreational uses and sports such as fishing, sailing and swimming. Reservoirs and dams, catchments, wetlands and marshes should also be described. Beaches, estuaries and harbour inlets should be represented clearly on maps, and their role and use by sections of the community fully detailed.

Back to Basics: The Components of an EIS

Some types of developments involving water trigger far-reaching changes of a physical nature. In coastal areas, mining and dredging activities can affect beach and harbour currents, tides and deposition of suspended materials. Wetlands and marshes are particularly vulnerable to developments along coastal margins. Recreational and commercial fishing may be jeopardised.

Attention should be paid to historical references to sudden changes in water flow, such as occur during a flashflood or one-in-a-hundred-year storm.

The distribution of water over a site also needs mention. This may alter during construction and after completion of a project. The quantity and quality of water present in one portion can increase and in another may decrease. Hydrological data is required detailing other sources of natural water, such as artesian and/or bore water, wells and aquifers.

Existing use of water on a project site should be noted together with site discharges of water and the quality of any discharge. These observations are important because the presence of existing pollutants will limit the potential of the site to handle additional materials.

Impacts of a physical nature occur on the water sector of the environment when rivers and watercourses are re-routed or, less visibly, as smaller streams and local runoff diverted around the site. Excavation and filling of an area will introduce new sites for water to accumulate, or change the depth of existing beds.

Leaching of minerals from soils can accompany mining and removal of topsoil. Re-vegetation may be delayed or prevented because of removal of essential nutrients. Animals which forage on and among plants migrate to other sites where food supply is better and more reliable. These are physical processes with chemical and biological implications.

Toxic pollutants, oils and heavy metals such as lead,

cadmium and mercury affect the quality of water and the survival of dependent living species. Residues of fertilisers, pesticides and their breakdown products readily enter local streams and groundwater as runoff from re-zoned agricultural lands. The resultant phosphates, nitrates, phenols, and dissolved salts can change groundwater chemistry and pH and subsequent treatment if intended for animal and human consumption. They provide nutrients for biological life, but within strict boundaries governed by a sensitive balance between toxicity resulting in death, tolerance and marginal survival.

Industrial processes and factory operation may also specify a need for water, e.g. hydroelectric power generation. Its use may result in changes in its temperature which can encourage increased growth of algae and microscopic life but at the expense of larger marine life including fish. Dissolved gases such as oxygen may be reduced by apparently minor changes to the balance which maintains the marine ecosystem.

Irrigation of semi-arid lands in an effort to improve agricultural value may involve construction of canals and re-routing of sections of a nearby river (physical effects). Excessive or imprudent use of the available water may result in raising of the water table with increased salting of the soil (chemical effect).

Clearly it is well-nigh impossible to divorce impacts and effects on the water sector (the hydrosphere) from other inter-related sectors of the ecosystem. We can seek to define the different contributing factors and processes, but in the end we need to view their contributions in conjunction with those of the geosphere, the hydrosphere and the biosphere. It is when we attempt to differentiate and delineate that this lesson is enforced, time and again.

The Soil

Developments affect the character of soil in two distinct

ways: first, through physical damage to soil structure, and second, through chemical damage caused by pollutants. Land-take for developments and construction are responsible for the loss of topsoil. Mining removes overburden and transfers it into dumps and tailings dams. Metals and minerals are selectively removed leaving debris and degraded materials, more susceptible to erosion and less capable of supporting the natural plant communities common to the area. Efforts to restore soil texture following construction are rarely successful.

Erosion is a frequent result of construction where vegetative cover, shelterbelts or forests have been cleared to permit a development to proceed. Heavy machinery used during construction may compact soil. Cut and fill, deep digging for foundations and piling may disrupt natural drainage. Waterlogging may occur in some areas: dessication of fertile soils denied water through re-direction of runoff into culverts and drains may follow motorway construction or shopping centre development. Alternatively, site drainage and runoff may be altered where concrete and building prevent runoff into soils and groundwater tables.

Direct pollution can result from deliberate dumping of wastes, emission fall-out, leaching and run-off. It upsets soil chemistry, pH and physical structure. Chemicals and wastes which affect plant life may denude the area of its vegetative cover and alter runoff patterns. Birds and animals may be poisoned by chemicals, or their food supply may be compromised as an indirect result of the pollutant. Residual chemicals may permeate the groundwater table.

Frequently, soils benefit from a development. Removal of wastes or minerals may greatly improve the quality of remaining soils. The visual appearance of the site is often enhanced by restorative measures including importing of topsoil, selective re-planting and care. Improvements to natural drainage can also result from a development.

An assessment of soil types should note soil type, structure and quality. Existing use, type of use and viability of associated industry needs to be made. For example, soil used for growing wheat varieties which carry high gluten (protein) are capable of generating maximal prices in sales and produce high export income.

Vulnerable soils are readily susceptible to erosion or damage; e.g. cleared, poorly covered grazing carrying large numbers of stock. These and other soil types may be protected or quarantined for conservation, or as catchment areas, heritage sites or as natural peat beds.

Flora

It is extremely difficult, if not impossible to separate the environmental effects on flora and fauna from each other or from their combined effects on the soil, air and water which are the basis of the natural ecosystem. It is, however, convenient to catalogue each sector separately even though the effects may be repeated in each.

The term "flora" refers to the vegetation which covers an area. These range from the smallest algae to the parasitic mistletoe, to embrace the oak, the elm, the eucalypt and the rose. Flora includes the bryophytes (mosses), ferns, lichens and cymbidium species (orchids), from the simplest gymnosperm (conifers) to the giant redwoods and complex flowering plants.

In the main, 'flora' are green plants - the principal producers in the ecosystem. By photosynthesis they capture light energy which they store in their cells as carbohydrates and simple sugars. Plant cells have walls which contain cellulose to provide structure: roots and leaves which connect them to, and facilitate their interaction with, the physical environment. They reproduce by both sexual and asexual means, by budding, pollination or cross-fertilisation. The environment either facilitates or frustrates these processes.

Any patch of land providing soil, water and light is

likely to support a myriad of plant life. These plants enhance the visual as well as the material value of the site. Other living species, principally animals and humans, depend on plants for food.

The botanist, ecologist or plant scientist preparing an assessment of flora for an environmental impact study will concentrate particularly on five key areas:

* rare species or those threatened by extinction: most countries contain a register of threatened or endangered species protected by legislation
* species which are essential components in an ecosystem, possibly because they fulfil a role in pollinating a number of other plants or one particularly vulnerable species. Interdependent relationships between species such as plants, between plants and insects, or plants and animals, are known as keystone mutuals and mobile links.
* commercial forests, natural rainforest areas, heritage and bushland areas
* plants, shrubs or trees acting as a visual asset to the landscape
* existing environmental effects which may already have put the flora at risk.

Re-zoning of land for settlement, deliberate or inadvertent vandalism and the removal of rare and indigenous plants are the most overt of the impacts which impinge on plant life. Access roads to construction sites, temporary waste and destructive incidental damage by workers during construction is likely to result in short and long-term loss of flora.

The destruction of large areas of forest, of shelterbelts and clearance of savannah is likely to affect the microclimate. Plant growth and health can be significantly affected by dust, grime and toxic emissions. Leaching from earthworks and tailings dams associated with mining sites can disturb the pH balance in the soil and result in plant loss.

Indirect effects on flora, both during and subsequent to completion of the project, include pollution of soil and water, changes to the water table as a result of clearance of vegetative cover, increased salinity and increased risk of erosion. Activities as diverse as mining, housing, controlled forestation logging, flooding of an area for hydroelectric purposes, irrigation and road construction will all affect prevalence and growth of flora.

It is sometimes difficult to see what benefits can outweigh potential adverse effects on local flora from development. A major planning proposal will usually highlight and galvanise public support for plants, flowers and trees, often from individuals who previously had tended to take such matters for granted.

Fauna

Fauna in an EIS includes all mammalian species, reptiles, rodents, birds, aquatic and insect life. Rare or threatened species need to be specially noted because they are protected by legislation. Any disturbances to the ecosystem and the relationships which fauna have with other parts and species in the system, will alter the delicate balance that often ensures survival of one or more marginal members.

Natural changes occasioned by weather or natural disasters such as earthquakes or landslip will, in themselves, result in changed relationships and species extinction. These play a time-governed role in the evolution and decline of new species. However, those activities resulting from human actions and changes are beyond the norm and are artificially imposed. They tend to occur in a very short time and often result in complete loss of one or more species, thereby disrupting the life chances of all others dependent on them. On occasions, human activities can result in the disappearance of a whole range of interdependent species in a given area through loss of one tiny but important member.

At the same time, some small animals, birds, insects, and species such as rats and mice, can tolerate and even thrive in a disturbed environment. Others, such as the larger carnivores, or exclusive herbivores like koala and fruit bats, may be more vulnerable. Scavengers such as the fox, hyena or dingo can gain benefits from humans, possibly through an improved food supply that can compensate, in part, for a loss of territory and increased risk of predation. Species classed as pests or vermin are mentioned in an EIA because of their role in the support of other species in the ecosystem. Insect species are noted with special attention to their functions as pollinators or in biological control of pests.

Migratory species, particularly birds, are important, and those species present should be detailed and their migratory patterns described. Coastal wetlands and riverine deltas, inland waterways and tidal lakes frequented by birds on a seasonal or regular basis (e.g. breeding, nesting) benefit from the additional wildlife that follows them, intent on predation.

Feeding of aquatic birds disturbs and redistributes deposits on the bottom of semi-stagnant shallow waters: birds' droppings add nutrients which dissolve to provide food for simple marine plants, small animals, and larger plants such as mangroves and aquatic weeds. These plants serve to filter out suspended and dissolved pollutants and maintain the water quality so essential to plankton, small arthropods and simple aquatic animals.

In most instances, an environmental effect which is adverse to one species is likely to be adverse to others, since prey and predators are closely linked and some species depend on a limited range of food for their survival. The most obvious direct effect is land-take, usually affecting the feeding territory, shelter or breeding sites of a species. Aquatic life is directly affected by changes in water chemistry due to discharges or run-off into streams, pollution or thermal discharges, destruction of bank habitat, silting or flooding due to changes in water management or changes to groundwater and catchment.

Indirect effects of population changes range from overfishing to excessive legal slaughter (e.g. kangaroos and wildfowl), disturbance of nesting sites and deliberate dumping of refuse or waste vegetation in streams or ponds. Vandalism and illegal hunting, pollution of water or food supply, and changes in water supply or location also need acknowledgement.

Increased awareness of native wildlife and concern for their conservation frequently accompanies the EIA process. The detailed listing of species by consultants is frequently followed by ongoing research. This often results in the identification and listing of new interdependent species, including dependent flora and micro-organisms. Society as a whole benefits when this occurs. The community further gains when new conservation areas are declared by a developer keen to gain public support for the project. On occasions, wildlife will further benefit, especially if feral species numbers and human predation are policed.

Human Beings

Human beings in an environment and their actions are considered, in the first instance, in an individual sense and in the second, as a community. Their actions similarly exert effects at a number of different levels. There are those which are of a social nature and involve income, work and status. Others affect basic welfare in terms of health, schooling, housing, food and safety.

The main human sectors which are likely to be subject to the most significant environmental effects focus on:
* the community: e.g. town, village, neighbourhood, people living in a 5-kilometre radius of the site
* demographics: e.g. numbers, ages, social classes, ethnicity, linguistic differences
* institutions: e.g. hospitals, residential enclaves (retire-

ment villages), colleges, half-way houses, prisons
* employment: e.g. unemployed, retired, students, skilled and unskilled workers, etc.
* education for subject community and wider community: e.g. primary, secondary and tertiary, part-time and full-time studies, technical and trade, etc.
* transport, private, public and communal - use of local roads and networks by local and through traffic, access to motorways, parking, rail links, access to stations, commuter patterns, bus services and usage
* shopping for local communities and wider catchment population - shopping centres, markets, regional and local, size and type of catchment areas
* welfare for the community - emergency services, health services, local authority services, voluntary services
* law and order maintenance - police numbers and location, level and type of crime, response to calls for aid
* recreation of all types - sports, passive recreation, entertainment, clubs, informal meeting halls

Table 2.5: Social indicator data categories
Statistical collections:
census data, household ownership, employment statistics, car ownership, crime rates, education participation rates, workforce statistics
Documentary data:
media reports, historical documents, government reports and policies, parliamentary Hansard
Observational data:
participant observation, action research, anecdotal evidence
Survey data:
interviews, surveys, polls, citizen responses, lodgements of objections, petitions

The availability and nature of employment, the numbers, gender, training and expertise of the employees, their needs, amenities and entertainment are almost always altered in some way by a development - major or minor. The assessment of the human environment therefore involves the collection of a vast amount of social data on the immediate community, often of a highly personal nature.

Existing stresses in the community should attract special mention. Issues such as housing shortage, shortage of rental accommodation, high unemployment, lack of adequate primary schools, etc., must all be given special mention. Other equally important but more sensitive issues also deserve mention because they may be areas in which an impact can occur. For example, evidence of an undercurrent of animosity between ethnic communities is likely to be aggravated if a particular development results in increasing numbers and improved facilities for one group at the expense of the other, thereby distorting the racial balance in the total community.

Valuable techniques for data collection for social impact assessment include:
* demographic analysis
* social surveys
* community studies
* individual case studies
* institutional analysis

Because of the eclectic nature of the above data, the collection process for social data is often prolonged and can involve dealing with a range of different disciplines and institutions. Accurate and reliable statistical data is required to substantiate discussion in all of the above areas.

At the same time, it needs to be remembered that social data always contains subjective elements and is "value-free". Nonetheless, it offers opportunities for a broader, more practical and realistic appreciation of actual effects, as felt, expressed and experienced by people themselves.

Combining the macro elements

The principal sectors of an EIS discussed above can also be described in terms of their component elements. For example, atmospheric effects, including the presence of natural or industrial pollutants, are important to considerations of climate, at the same time as they have an impact on living species - humans, animals and plants.

Depending on the location of a project, the atmosphere may be of major or minimal importance. Communities located near to a new development may receive significant amounts of pollutants. Some projects have specific requirements for clean air and isolated environments. Pollutants in the atmosphere may also cross national and continental boundaries. Emissions, radiation and dust fallout from volcanic eruptions can travel around the globe, as shown by the 1991 eruption of Mt.Pinatubo in the Phillipines.

Economic issues and implications also need to be identified in an EIS. Fundamentally, economists assess the value of a project on the basis of its relative cost and predicted benefits. In establishing the cost of a project, attention needs to be paid to the environment which is altered by the project. This focuses on social and political factors and sometimes raises difficult philosophical questions, value issues and aesthetics.

In a commercial context, it is crucial that the project generate monetary profits as much and probably more than that it should carry social benefits such as increased employment and improvements in community income. Relevant are factors such as the increased income generated in a local community by a new industrial or mining venture. Increased migration into the area, a higher level skills base among the local people, improvements in housing, sanitation, drainage and transport follow, together with more schools, more amenities, better health services and an improvement in the quality of life.

In depressed communities and those where unemployment is high and facilities poor, these gains will often far outweigh the loss of high quality agricultural land and beauty spots. The possible extinction or marginalisation of some plant and animal life and the costs associated with dealing with the wastes arising from development, are balanced against the financial and social prospects.

The issue of relative cost, benefits and balance however, raises a very real problem and one which enters into any EIS. Costing is that part of the auditing process which follows the planning, descriptive and detailing phases considered earlier. Its role in the EIA process is more fully addressed in the following Chapter outlining the steps which are taken by developers and consultants in preparing an EIS.

Perhaps the easiest way to come to appreciate the difficulties in costing items in an EIS is by way of examples. How does one establish a 'true' value or price to a natural resource? Is a non-renewable resource (such as a mineral deposit) a capital asset which cannot be used to produce income without loss of that asset? How does one decide whether the product is of more or lesser value than the resource from which it was created?

What price can be placed on the wind-swept desert of central Australia, rich in old ore and bauxite? Unmined, unused and unappreciated, the desert may not appear to carry much value. It could be presumed to be an asset, but one of an intangible nature. Mined, the same desert could yield vast incomes which could materially improve the wellbeing of all the Australian community, a tangible asset. Similar arguments might be tendered for projects seeking to drill the tropical wetlands of the Malay peninsula believed to hold vast oil reserves beneath the mangroves.

Aesthetic values and questions expose other anomalies and priorities. For some, the Great Barrier Reef off northern Australia and the high mountains of the Himalayas are areas of such untold natural beauty that the merest sign of human intrusion is a travesty. Others prefer the hum of activity and the thrill of tense-packed urban life, with a constant bustle of people and an ever-improving diversity of new products, new activities, new life opportunities.

The pricing of visual landscapes, heritage and cultural sites (intangible assets) is probably among the most difficult for the environmental economist. Individual tastes and preferences vary. Can abstract ideas and enjoyable environments ever be costed, in traditional monetary terms?

Other significant development-related effects which need to be noted include:
* gains and losses in property value
* increases in transport costs versus improved public transport
* gains and losses in trading income
* compulsory land acquisitions
* archaeological and cultural sites identified during the project and its construction phase
* noise, nuisance and traffic hazards

Finally, wastes and residues accompany any construction and work activity, and they must be acknowledged in the EIS. Their nature - chemical, physical or biological - and their quantity and type should be noted. Most important is the ability of the material to degrade or break down naturally and its potential for recycling. Different classes of wastes may all be produced in any one site or development: organic solids, tailings from mine workings, gaseous emissions, oils and other petroleum products, phenyls and byproducts of complex chemical processes.

Intractable wastes present on the site prior to development need to be identified and their remediation costed in realistic and practical terms. Heavy metals such as lead and arsenic, nuclear wastes and organochlorine chemicals pose special management problems if produced by the development under consideration. Their presence on a site means that costly remedial measures may be called for; for example, removal, transport and storage of soils contaminated with arsenic from cattle tick dips which need to be secured in concrete bunkers. Alternatively, wastes produced during the manufacturing processes associated with a proposed development need to be handled, ideally on site or transfered elsewhere. Both measures involve costs which need to be factored into the EIS as well as the project tender price. In some instances, new technologies will permit exploitation of the waste and this can serve as a secondary revenue source for the new project. Unfortunately, technologies for waste management are still in their infancy and the opportunities represented by them are yet to be communicated to business and the wider community.

In the matter of human and animal health, many of the conditions blamed on pollutants and wastes arising from a development, together with social dislocation and psychological effects, are poorly understood. Cancers take years to develop: birth defects and changes to genetic heritage occur over the generations. Today, health professionals and researchers are closely monitoring changes to human health, in order to identify ever more early warning symptoms and evidence of toxicity which can be confidently linked to industrial and other developments.

At the same time, human and animal species continue to possess a remarkable capacity for adaptation. Developments may, in fact, serve to benefit a species by acting as an agent for natural selection, thereby permitting a more robust species to survive, better suited to meet the challenges of the new environment. While initially resistant to change, human individuals tend to adjust socially and psychologically to new circumstances in a relatively short period of time. And, there are also the mitigating effects to be considered.

Concluding Remarks

The whole scenario, however, suggests that we are only just

beginning to appreciate fully the implications of human activities for planet earth. It would appear that until just recently, humans exploited this planet (along with other species) and forged a mutually beneficial and productive relationship. Now there are just too many human exploiters to maintain the same balance.

As we have seen above and observe in our own lives, times change, as do values and priorities. The EIA process offers an important opportunity for humans to think twice before launching into ill-planned ventures and pursuing ambitious, greedy schemes for personal profit.

CHAPTER 3:
FRAMEWORK FOR ENVIRONMENTAL IMPACT STUDY:
The seven stage system

Chapter 3 takes the basic data discussed in Chapter 2 and looks at where and how it fits into the formal environmental impact assessment process. It explains how decisions are made about whether or not a project is subject to an EIS, and then the approach which the developer must take to prepare a comprehensive impact assessment in order to obtain approval from the relevant authorities.

Fundamentally, the EIA process consists of seven distinct stages:

1. Screening	Does this project need an EIS?
2. Scoping	What issues should be addressed in the EIS?
3. Profiling	What is the current status of the environment in the affected area?
4. Risk Assessment	What are the risks or benefits? Who is likely to be affected?
5. Risk Management	Can risk be avoided or prevented?

Are better alternatives available?
How can benefits and risks be costed?
How best can benefits and costs be negotiated?

6. **Implementation and Decision-Making** Does assessment provide adequate, valid, reliable data suitable for decision making?
Is there a conflict to be resolved?
How will conditions be enforced?
How and by whom will impacts be monitored?

7. **Monitoring and Post-Project Evaluation**
Following completion, is the project complying with its conditions: are the anticipated outcomes being achieved?
Are protections embraced adequate or deficient?

The following summary details of this chapter borrow heavily from material prepared during the development of a national framework for health impact assessment for the Australian Government and coordinated by a small four-member team (Ewart, Young, Bryant and Calvert, 1992). Their work is still under discussion and awaits formal recognition by the government.

Screening

Screening aims to identify developments and projects that should be subject to an EIS. These days, this process is assisted by the existence of laws specifying those industries and types of developments which are likely to have an impact on the existing environment.

Screening is usually carried out by a government authority or department, e.g. Department of Planning, Lands or Housing. However, the nature of the assessment process, types of issues to be canvassed and the way in which it is conducted are not usually detailed. There is often no guarantee that all the relevant and interested authorities have been approached, or that public concerns such as short and long-term health impacts have been sought or addressed.

Table 3.1: Components of screening
* rapid assessment checklist for criteria which establish
   need for EIA
* local demographic data - identify vulnerable human
   communities
* local environmental data identify vulnerable ecosystems
* assessment of potential global impact of proposals
* rapid assessment of public interest and/or concern
* communication networks between community and
   developer
* relevant intersectoral policies

Scoping

The USA National Environmental Policy Act describes scoping as the mandatory use of mediation to involve interested parties in the design of an environmental assessment. Scoping identifies the issues to be considered in depth and those to be eliminated: it allocates responsibilities for researching and preparing the evidence.

Key tasks in the scoping process include:
* identifying key stakeholders, including community
   groups and public instrumentalities.
* determination of the feasibility of alternatives and the
   need for further consideration of such alternatives.
* provision of guidelines for the proponents and specifying information about:
      (i) goals for impact assessment and criteria for

monitoring
(ii) minimum information requirements
(iii) nature and location of relevant existing social
     and health data concerning affected populations
(iv) relevant bio-technical standards and inventories
(v) relevant psycho-social and health indicators
(vi) relevant biological and social assessment
     methodologies
(vii) recommended consultation strategies
(viii) recommended post-project evaluation and
     monitoring protocols
(ix) local agency requirements and policies.

Baseline studies of the existing environment are required for inclusion in any comprehensive EIA. At the very least they should include population, housing settlement, land use, meteorological and biological surveys. There is a need for studies of social status, occupations, health, ethnicity, age and skills possessed by the affected people. It follows that people living and working in the environment need to be informed of project proposals at an early date. They need to be encouraged to cooperate at all stages in the data collection process.

Profiling

Profiling is the process of establishing the baseline condition of relevant parameters of an affected community so that likely impacts can be predicted and subsequently monitored. Some of this data may already have been collected during the screening and scoping process. The accurate and detailed collection of baseline data is essential to ensure that actual changes or effects resulting from a proposal can be accurately studied and quantified. A summary of the necessary baseline data is contained in Table 3.2.

Table 3.2. Baseline information (profiling)
Characteristics of the existing and incoming populations of the region including size, age structure, socio-economic status and groups at risk.
Physical characteristics of the region such as frequency of atmospheric inversions, variability of river flow and orientation of prevailing winds.

Existing land uses, especially those that can be considered incompatible or inappropriate.

Current health status of the population including morbidity and mortality characteristics, known pathways for existing diseases.

Current levels of pollutants and environmental quality.

Existing data and studies concerning types of problems likely to arise from the development, such as existing criteria for air or water quality, and known relationships between removal of native vegetation and native species, pollutants and human health.

Existing standards of living of the population, especially in relation to factors such as access to water supplies, adequacy of diet and accesss to health facilities.

Baseline information gathered should include analyses of the direct and indirect implications of a proposal. Impacts on local human, biological and wildlife populations should then be monitored in an holistic respect. Identification of potential impacts, follow-up evaluation and monitoring and the analysis of the data which result, depend also on the existence of indicators or criteria which define the range of outcomes.

The Use of Indicators

Indicators become important where impacts are likely to be cumulative, complex and multifactorial. Some general principles should apply to inclusion of the following in any environmental impact statement (EIS):

1. Low impact effects develop over time, have a long latency period, are difficult to measure or are evidenced only under stress conditions, e.g. low-level exposure to potentially carcinogenic herbicides; these require inclusion.

2. The EIS must allow for baseline data which is incomplete, of limited reliability or difficult to obtain.

3. Historical data covering exposure to a wide range of criteria - even those judged "safe" today - may in future be judged detrimental. Biological markers of exposure need to be identified and monitored to indicate where more in-depth quantitative risk assessment is needed.

4. Assessment of likely exposure is the first essential step in quantitative risk assessment. In the biological and chemical field, dispersion modelling and pathway analysis will involve the chemical, ecological and geographical studies of the distribution of toxic substances in air, ground and water as well as studies of the habits and behaviour of people in the vicinity. This might determine their level of exposure.

Specific health-related data include:
* morbidity, and mortality data
* birth and infant mortality rates
* specific disease rates
* epidemiological surveillance data
* indicator disease registers such as cancer and asthma
* health service utilisation and
* health expenditure

Risk Analysis

Risk analysis is the primary tool for bringing quantitative and semi-quantitative data to bear in the decision making processes associated with hazard identification and mitigation.

Risk assessment involves the scientific identification and quantification of environmental hazards and the relative risks of various options, based on available evidence

and experience, and summarised as:
* hazard identification
* risk estimation
* risk characterisation
    The risk assessment process is summarised in Table 3.3.

Table 3.3: Types of environmental impact assessments
1. assessment of new development policy or program proposals
2. cumulative regional impacts
3. clean-up/de-contamination of disused sites prior to reversion to other uses
4. general on-going environmental audit and monitoring
5. licensing and control procedures - requirements

    Risk management where decision makers and interest groups develop policy, guidelines or management procedures based on the results of risk assessment combined with feasibility, and economic and socio-political realities. These can be characterised as:
* risk communication
* setting standards
* mitigation measures
    Therefore, risk analysis is the most important and time-consuming component of the EIS. Relevant scientific information not only needs to be collected but condensed into a manageable form. Unfortunately, time constraints are imposed on data collection. For any one parameter, this means that the consultant responsible reviews existing conditions in a given site, traces the history of previous changes and results, and where data is available, makes a comparison with similar sites exposed to the environmental challenges similar to those planned.
    For example, a project meteorologist would be required to undertake a comprehensive survey of normal rainfall patterns, atmospheric conditions, temperature variations, wind and seasonal climate variables. Extremes such as drought and flood, thunderstorm activity, frequency and

severity would be reviewed where records are available. On site recording and monitoring would provide recent substantive data to support historical observations, where appropriate, with regard to microclimatic variability associated with topographical differences over the project site.

This data would be summarised and combined with data from other specialists on the scientific consultancy panel. The membership of these panels and the weighting or priority given each member's contribution is dictated by the nature of the project. These matters, together with an outline of panel membership is discussed more fully in the following Chapter 4.

A further issue which arises in the assessment process is the difficulty of finding suitable analytical models which accommodate the complex environmental variables involved, together with various alternatives and their outcomes. Computers have greatly simplified and upgraded the analysis of EIS data. Sophisticated software is now tailored to specific types of EIA and determined by the nature of the project under consideration. The quality of predictions and the range of impacts which can be simultaneously assessed minimise the potential for poor judgement and subjective bias. This places increased importance on the data collection process.

Implementation and Decision Making

In EIA, decision makers such as local councils, government agencies, and commercial developers are usually faced with a number of options regarding any particular proposal. Their goals will be:
* implementation of the project as proposed
* the design of measures to be incorporated into the proposals to
    (i) optimise the potential benefits of the proposal and
    (ii) prevent or minimise undesirable effects (miti

gating measures).

If objections to the original proposal warrant, the decision makers may adopt an alternative development which achieves the same basic objectives with less impact on the environment. Cost and political considerations may play a heavy role in any final decision whether or not to proceed with the project.

For urban developments in particular, choices are not always clear. The development which eventuates may be a compromise between a number of options, each with certain advantages and disadvantages. In many circumstances there are no alternatives except to proceed (with or without mitigation measures) or to abandon the project.

Table 3.4 summarises the implementation and decision making activities phase of the EIA process.

Table 3.4. Risk assessment

HAZARD IDENTIFICATION: Is there sufficient evidence to suggest that a hazard will occur?
* analysis of properties of the development
* epidemiological evidence of risk
* toxicological evidence of risk
* case reports

RISK ESTIMATION: What is the likely extent of risk?
* impact on environment
* dose-response and effect relationships
* population exposure and sensitivity
* probability and consequences of accidental events

RISK CHARACTERISATION: Is it the best option, worth the risk?
* public perception of risk
* risk-benefit analysis
* feasibility of alternatives
* social, political and cultural implications of each

* acknowledge uncertainties and options
* social justice and equity considerations
Source: after Ewan *et al*, 1992, p55

Decision Making Processes: Toxic Waste Facility

A recent example comes from Australia, which has a growing problem in disposing of intractable toxic wastes. This can be defined as 'any waste that does not break down naturally and for which there is no environmentally acceptable means of disposal'.

Table 3.5. Implementation and decision making
* impact assessment must provide sufficient, valid and reliable data for decision makers
* implementation options and decisions should be clear and give adequate opportunity for input by all stakeholders
* dispute mediation and mechanisms for handling compensation claims should be provided for
* enforcement mechanisms to ensure compliance with development conditions, including linkage between routine monitoring and re-registration

The only recognised method of dealing with such wastes has been to burn it at very high temperatures. Yet questions remain regarding the health risks to workers and those living near the facilities. A number of European countries, notably France and Germany, possess commercial incinerators which accept these wastes. Australia does not have a high-temperature incinerator, and until recently sent such wastes for disposal to Wales. A proposal to build a high-temperature facility in Corowa, rural New South Wales had received government approval to proceed in 1990. The local community was advised to look forward to many new jobs and employment opportunities which indirectly would result in increased importance for their small town.

Unfortunately for the planners, the prospect of being the site of a high-temperature waste disposal facility resulted in a massive public outcry and generated considerable publicity. Eventually the project was shelved - to be sited elsewhere. However, local populations in alternative sites were equally vehemently opposed to the project. As a result the incinerator was abandoned. The intractable waste problem was placed in the hands of a specially convened panel. Over a two-year period, the panel was commissioned to investigate methods and ways that Australia could manage its toxic wastes into the twenty-first century.

The Independent Panel on Intractable Waste submitted its report in November 1992. The Panel found that a high-temperature incinerator was not only unpopular but expensive to build, difficult to run and could not guarantee a period of operation to justify expenses in establishment. Rather than opting for a single location and temporary storage of wastes prior to incineration, the Panel recommended a suite of smaller technologies currently under development and which permit disposal of wastes at the site where they are generated.

These recommendations acknowledge the fact that any technology appropriate for disposing of polychlorinated biphenyls (PCBs) may not be equally successful in disposing of low level nuclear wastes (radionucleotides), contaminated hospital wastes (biological) or dioxin residues in paper manufacture (organochlorines). If they are accepted by the Australian government, the responsibility lies with the government to provide adequate funding to encourage technologically innovative companies and scientists to develop these new avenues for waste disposal. Initial work has yielded some promising prospects, including a mobile unit able to handle a wide range of materials. A number of permanent and semi-permanent storage sites for toxic materials have also been designated.

In the intervening period and until the new technologies are refined, Australia must continue to export its wastes, at considerable expense, to countries with appropriate facilities to dispose of them safely (*Sydney Morning Herald*, 16 November 1992: Report of Panel on Intractable Wastes, November 1992, AGPS).

Environmental Auditing

Environmental auditing measures the extent to which a project is meeting its environmental, social, economic and health goals. In other words, auditing refers to the formal means by which the performance of an environmental management program can be evaluated and adjusted.

In most cases, environmental auditing is performed to verify compliance. Increasingly, however, large corporations and smaller concerns are using auditing for public relations purposes to promote the image of the good corporate citizen. ICI, for example, conducts regular environmental audits. The most recent report, released in 1992, shows that ICI is achieving profit and growth while preserving the environment or acting to reduce, eliminate or minimise the costs and damages associated with its activities.

Environmental audit is defined by the Confederation of British Industry as 'the systematic examination of the interactions between any business operation and its surroundings'. This includes all emissions to air, land and water; legal constraints; the effects on the neighbouring community, landscape and ecology; and the public's perception of the operating company in the local area.

Company conducted audits usually contain actual figures and statistics which allow data to be contrasted with industry averages. This permits monitoring of progress made towards environmental targets defined for other companies with comparable operations. These benchmark comparisons add credibility to a company audit. A 'true' audit of achievements in environmental conservation also includes a complete breakdown of all hazardous waste, for example, not just data related to a single pollutant such as chlorofluoro-carbons (CFCs) produced.

At times, and in order to limit public concern at high levels of emissions or waste produced, companies refer to the size of their operations. Reduced levels of production or 'down-sizing' of a company can often be forgotten unless emission reductions are compared with changes in productive output. It is possible that greater efficiency and lower levels of toxic emissions can be achieved at high levels of output during a manufacturing process, e.g. blast furnace in steel production.

Data also needs explanation. In *New Scientist*, 3 October 1992, pp.21-2, Pearce suggests that, while British Airways contributed but a mere 2 per cent to worldwide airline emissions of carbon dioxide and sulphur dioxide in 1991-2, its carbon monoxide emissions top 4.5 per cent and nitrogen oxide emissions are 2.5 per cent, well in excess of BA's proportion of 2 per cent passenger mileage (British Airways Environmental Report, 1992). Nor are reasons given to explain these inconsistencies.

Environmental audits also refer to company targets. These confirm environmental protection as company policy. They guarantee commitment to defined goals and enable progress towards them to be gauged. Historical data permits direct comparisons between companies on progress towards achieving these goals.

In Britain, the Chartered Association of Certified Accounts (CACA) has been working on proposals for common standards in green auditing so that outsiders can make sensible comparisons between companies. Complete inventories and statistics on a company's operations are necessary for individuals to be able to compare actual performances and the environmental targets of competing firms.

Companies differ widely in their interpretation of auditing processes. Some use auditing as an excuse to promote unpopular staff cuts and technological changes on the factory floor, usually by reference to a commitment or a target to meeting national goals for environmental protection. For example, claims that a changed procedure may lead to reduction in greenhouse gas emissions even if it may have cost jobs, causes major disruption and restructuring of the workforce or increased costs to consumers.

Finally, data can be presented in a misleading manner and appear to view a company's performance in more positive terms than may be warranted. In 1992, British Airways reported that its aircraft emitted 11 million tonnes of carbon dioxide into the atmosphere. It fails to state, however, that this represents but less than 2 per cent of the total airline traffic about the globe.

Further Considerations in the EIA Process

There are a number of issues and players in the EIA process to which special attention needs to be directed. These include:

1. consideration of who can be defined as an 'interested party' to a development proposal;
2. an examination of the means or principles by which individuals decide whether a recognised risk is or is not acceptable, and under what terms;
3. attention to the means and techniques used to analyse risk.

Stakeholders

Five groups are concerned with the outcomes of EIA. Groups involved include:
* proponents of the projects
* decision makers at all levels of government
* practitioners and health professionals involved in developing the EIA
* individuals responsible for planning and service provision in post-project environment

* community members and workers who will be affected
  by the development, both as workers and residents in
  the affected area.

Each group will have a different concern, different
needs for information and differing criteria for evaluating
outcomes. Toxicologists, for example, may be concerned
about toxic emissions from a major mineral processing plant.
Workers, on the other hand may be more concerned about
personal safety and injury protection. Health professionals
may focus on accident and emergency facilities. Local residents may be worried about the impact of a 24-hour industrial operation on nocturnal noise levels and heavy vehicle
traffic.

The Public Acceptance of Risk

Public acceptance of risk depends firstly on the nature of
the risk and its level. It is also dependent on cultural, social
and psychological factors. Public attitudes diverge widely
and can be influenced, at the personal level, by an individual's level of education (e.g. scientist versus shop assistant),
and by one's socio-economic status (e.g. where the only work
for a poor man is in a factory producing chemicals, where
workers are highly likely to develop terminal cancers).
Unfortunately also, risk predictions are usually described
as statistical probabilities: a format difficult for average individuals to comprehend.

In general, people are less likely to accept a risk which
is involuntary, unfamiliar or potentially catastrophic or
which results in particularly emotive outcomes such as child
deaths. In democratic countries, public perception of risks
has practical as well as social implications since public agencies and legislation are driven by public concern. Environmental laws are shaped by these perceptions as much as,
and even more than, scientific understanding of risk. This
can result in resources of environmental, welfare and health

agencies being directed towards those environmental problems perceived as most serious by a local community, and often at the expense of other more pressing problems.

The mass media should be taking a greater responsibility in correctly and objectively educating the public about the true - rather than just emotive - risks associated with particular projects and developments. The siting of toxic waste disposal facilities near human settlements previously referred to in Australia, is an excellent example of the problem where scientific evidence of the low level of risk or danger associated with high temperature facilities constructed in accordance with recognised safety standards, failed to overcome emotive public opposition.

Analytical Models

Risk analysis is the technical component of the EIS. In the final analysis, the quality and reliability of the EIA will depend on the quality and quantity of the scientific data collected and the manner in which it is interpreted.

There are a number of techniques used to quantify or measure risk.
* the Leopold matrix, a complex mathematical technique involving as many as 8800 separate analytical components
* aggregation, sometimes called weighting and scaling
* overlay techniques
* adaptive environmental assessment and management (AEAM), a simulation based approach.

The aggregation method involves combining numerical values indicating individual impacts into an artificial model which reflects overall impact. The most popularly adopted system assigns value functions to individual environmental parameters such as tree cover, soil status, species diversity. The 'grade' or 'quality' of the environment can be judged by use of the same scale, e.g. percentage.

Each parameter is also ascribed a weight according to its putative importance. Environmental quality scores are multiplied by the appropriate weightings and added to give a total score of environmental quality for each option under consideration.

Unfortunately this method is highly subjective. The consultant makes a number of assumptions based on his/her own subjective judgement. Where a single consultant is involved, and the data is inconsistent or sparse, the risk is that the resulting recommendations may be biased and unreliable.

Another popular method is the overlay technique, once common to town planning and land use decisions, and one of the early methods used in environmental planning. It involves the production of transparencies showing the spatial distribution and intensity of individual impacts. These can be overlain successfully, one on top of the other.

At any one time, overlays can show environmental impacts from a wide mix of data sources; e.g. ecological sites, historic sites, visual, health, settlement, noise, severance and water measurements. Where large sites and a number of transparencies are involved, photography and hierarchical clustering (multiple overlay) can be used. These techniques are especially suitable for assessing landform conditions, population distribution, housing and farming settlements, etc., over a total site. They are valuable when decisions about establishment of a new satellite suburb in a formerly rural enclave are being made, because a vast amount of data and a wide range of different impacts are to be considered.

The analysis of risk often bedazzles even the most well educated non-mathematician. Inevitably it focuses on the balancing of risk and benefits associated with any two or more choices, and hangs on the question of probabilities.

The use of computers simplifies these analyses and also

removes or minimises the subjective component of the process.

Combining and Computing the EIS Data

Computer developments have vastly simplified and compressed the analytical process. Raw data files can be manipulated by changing weighting values or by aggregating types of impact in various combinations.

Today in the 1990s, a comprehensive range of software is available for undertaking environmental assessment tailored to specific mixes of parameters deemed to be relevant to the proposal under consideration. The screening and scoping elements of the EIA now serve as a guide to the software requirements for the subsequent management of the data.

Computer simulations further extend the predictive power available to the investigator.

Final Remarks

For many, the above discussion will seem over-powering. Most people readily admit that they had no true appreciation of the complexity of the EIA process and the extensive nature of data collection and analysis involved. However, for the average suburbanite simply wanting to make a private submission to a public enquiry concerning, say, a proposal to claim an area of local reserve, bushland or forest for housing or to build a prison, the above seven steps should act as a guide rather than a comprehensive checklist.

The following Chapter uses some simple examples of two sections of the EIS which most interest members of the public: impacts on flora and fauna. Technical data, reference to sophisticated survey techniques and analysis are avoided.

CHAPTER 4
COLLECTING THE DATA:
What's involved

In the previous Chapter we learned how to put together the various component information for an environmental impact statement. Now we turn to the matter of Field Study - the manner in which information is collected and described by various scientists and experts who act as consultants. These are highly technical areas, and here the intention is merely to touch the surface to enable the reader to gain some insight into the details required.

In the first section of the chapter, the focus is on the plants and animals in an area. It aims to show you how to go about studying plant and animal communities in the field.

This is the type of basic exercise you need to be able to do in order to collect useful data for any public comment on populations and living bio-communities which are likely to be adversely affected by a development in your area.

The Study of Plant Communities
Plant communities are normally the first aspect of the environment studied by field biologists. This arises for a number

of reasons:

(i) the plant community comprising primary producers, represents the primary energy source upon which the whole ecosystem depends. It is green plants which capture and convert light energy by photosynthesis into the essential carbon-based sugars vital for life and growth.

(ii) the plant community is normally responsible for providing the shelter and breeding sites for other living populations such as micro-organisms and animals.

(iii) plant communities are strongly influenced by the prevailing climatic conditions, type of soil and availability of nutrients: as such they provide a ready source for monitoring changes in these physical features of the environment, as experienced by the plants themselves and other dependent species.

(iv) unlike the animal community, plant communities are sedentary and more readily observable than members of the detritus cycle.

To the observer, plant communities represent a bewildering array of diverse forms. Ecologists traditionally draw boundaries and may appear to make arbitrary decisions in an attempt to classify and describe the widely differing communities observed.

The first approach to the classification of plant communities is a structural one based on recognising the dominant *life forms* in the community. The *life form* defines whether the plant is a tree, a shrub or an herb.

To explain these groupings further:

A *tree* is a woody plant, more than 5 metres tall, usually with a single stem.

A *shrub* is also a woody plant, less than 8 metres tall and usually with many stems arising at (or near) its base.

An *herb* refers to the rest of the plant community which lacks secondary growth (that is woody, growth). This classification of a herb includes mosses, ferns, grasses and those plants found in streams, swamps and at the foot of trees or among shrubs, and whose flowers provide so much of the beauty of the landscape. In more detailed studies, distinctions are made between each of these types of herb.

The tallest form of life found in a community is regarded as being the 'dominant' life form.

The next stage in the classification is to estimate the extent of ground coverage by the dominant life forms. Four divisions are generally used: dense (70 to 100 per cent ground coverage), mid-dense (30 to 70 per cent), sparse (10 to 30 per cent) and very sparse (less than 10 per cent).

The percentage area of coverage of the ground by trees, for example, refers to the average area of ground shaded by the branches of the tree, the canopy or the area occupied by branches and trees: not that covered merely by tree trunks. This is because the area of ground shaded by the canopy has important implications for the growth of other plant life forms and dependent micro-organisms. The more open the canopy, the greater the opportunity for growth and development of shrubs and herbs and the survival of an increasing diversity of wildlife species.

These four basic divisions permit definition of the basic structural forms of most vegetation in the environment. Three basic categories are common: communities dominated by trees, communities where shrubs are dominant, and communities in which herbs dominate.

While recognising that each structural unit can vary widely in specific composition, it is possible to identify in general terms the common species normally associated with each grouping.

For example, dense forests are typically called rain-forests. They possess dense canopies in the upper stratum so that little sunlight penetrates to lower strata, thus limiting plant growth on the forest floor. In terms of species composition these forests fall into two groups: the so-called

Indo-Malaysian elements and the Antarctic (or Arctic) assemblage. Typically the former has an ever-green upper stratum, though some deciduous trees are found in drier areas. A great variety of the trees found represent angiosperm families. Lianas (vines) are invariably present.

In contrast, the Antarctic (or Arctic) assemblage is characterised by very few trees in the upper canopy. Lianas are absent, as are the understorey and lower tree layers. The dominant tree is likely to be a member of the beech family such as *Nothofogus*. Tree trunks are often covered with epiphytic mosses and ferns. Mixtures may occur with sclerophyllous, xeromorphic and eucalypt varieties, depending on latitude.

Trees are further described by their crown depth (the portion of the trunk from which the branches arise) and the bole (the portion of the trunk between the ground and the first branch). Early research labelled forests on the basis of the understorey of shrubs and herbaceous ground stratum. Usually the dominant tree species and structure of the understorey differ.

This can be illustrated quite simply by examining the nature of the understorey in soils of differing fertility. In infertile soils, for example, the understorey is dominated by xeromorphic shrubs with little herbaceous ground stratum. This ranges to an understorey dominated by herbs on fertile soils.

Other observations can be made based on the nature of the forest. Low open-forests are typically found in stressed environments such as very dry areas or near the tree line in alpine climes. Fire changes the specific composition of the plant community: it may arise on a regular or only occasional basis, such as may result from a lightning strike in very dry conditions following drying off of lush undergrowth. Even where ground fires occur periodically, they rarely destroy the dominant trees. Unless deliberate clearance and firing has been undertaken by human beings, the character of the landscape and treescape remains unchanged. In the past, human actions have contributed to the desertification of Australia, the relentless march of the desert currently occurring in Niger, Nigeria, Mali and other countries of central Africa, and more recently, the burning of large tracts of cleared forests in Malaysia and Brazil.

The following three Tables describe the structural characteristics of plant communities dominated by trees, shrubs and herbs.

Table 4.1: The structural characteristics of plant communities dominated by trees

HEIGHT OF TALLEST TREES

Coverage by tallest trees	*more than 30m*	*10-30m*	*5-10m*
Dense (70-100%)	tall closed-forest	closed-forest	low closed-forest
Mid-dense (30-70%)	tall open-forest	open-forest	low open-forest
Sparse (10-30%)	tall woodland	woodland	low woodland
Very sparse (less 10%)	tall open-woodland	open-woodland	low open-woodland

Table 4.2: The structural characteristics of plant communities dominated by shrubs

HEIGHT OF TALLEST SHRUBS

Coverage by shrubs	*2 to 8 metres*	*0 to 2 metres*
Dense (70-100%)	closed shrub	closed heath
Mid-dense (30-70%)	open shrub	open heath

Sparse (10-30%)	tall shrubland	low shrubland
Very sparse (less 10%)	tall open shrubland	low open shrubland

Table 4.3: The structural characteristics of plant communities dominated by herbs

LIFE FORM	COVERAGE BY HERBS		
	DENSE	MID-DENSE	SPARSE
Hummock grasses (0-2m)			hummock grassland
Herbs (various)	closed herbland	herbland	open herbland
	closed tussock grassland	tussock grassland	open tussock grassland
	closed grassland	grassland	open grassland
	closed herbfield	herbfield	open herbfield
	closed sedgeland	sedgeland	open sedgeland
	closed fernland	fernland	open fernland
	closed mossland	mossland	open mossland

The classification of 'Very sparse' very rarely exists.

A typical description of the plant community in an area for inclusion in an EIS follows.

EXAMPLE: *Hypothetical Study of Plant Communities, Wollondilly River, Panagonia.*

A transect at least 10 metres long and preferably down a slope is taken, and a profile drawn to scale.The profile should reflect the vertical as well as the horizontal features of the area and should compare such features as tree and shrub height, crown and bole height, canopy spread and shape, etc. Ferns, rocks, streams, aquatic weeds and prominent ant or animal mounds should also be represented.

Structure of the Community

The area from the northern edge of the transect (see Figure 5.1) to about 40 metres south is tall open forest dominated by trees reaching heights of 40 metres with bole height much greater than that of the canopy. A second storey has started to develop with scattered trees to a height of 6 metres with canopy height exceeding bole height. The plants in this section of the transect are growing in well drained sandy loam.

The section of the transect from the 40-metre mark to the 100-metre mark is on an incline and dominated by rocky sandstone and poor soil. The open forest is occupied mainly by tree species to 15 metres with bole height greater than canopy height. A second layer of smaller trees grow to 8 metres with similar bole and canopy heights. The ground is covered by diverse shrubs, scattered herbs and grasses with moss growing on the southern side of the rocks where there is little penetration by sunlight.

Description of Plant Life Forms

(i) <u>Tree Life Forms</u>: A variety of eucalyptus species forest tree life forms occur along the transect including: blue gum *(E. deanei)*, grey gum *(E. punctata),* yellow blood wood *(E. eximia)* and Sydney peppermint *(E.piperita)* which grow to respective heights of 40 metres, 30 metres, 30 metres and 20 metres.

(ii) <u>Lithophyte Growth Forms</u>: orchids can be found growing on rocks including species of Dendrobium; more occur on trees *(Cymbidium spp)*. Bromiliads are scattered in-between loose rocks along the southern side of the transect.

Collecting the Data

(iii) <u>Tree Fern Growth Forms</u>: many ferns grow in the damp valley floors, on the underside of rocks and along the banks of the creek. Examples include: King fern *(Todea barbara)*, tree fern *(Cyatheas australia)*, pouched coral fern *(Gleichenia dicaeca)*, umbrella fern *(Sticherus spp.)* and maidenhair fern *(Adianthum aethiopicum)*. Small trees of the Acacia, Hakea and Casuarina species grow under the main canopy of eucalypts in the rocky surrounds above the creek floor plain.

(iv) <u>Herb Life Forms</u>: grasses and herbs are found on the ground in all parts of the transect but more individual species and healthier specimens are present in the damper areas of the floor and scattered amongst the fern species. Aquatic plants including common ribbon weed *(Vallisneria gigantea)* are found close to and in the margins of the creek.

(v) <u>Xeromorphic Shrub Life Forms</u>: Many shrubs, often particularly small leafed varieties with waxy cuticle and specific adaptations for water conservation, are found beneath smaller tree species; examples include *Bossiaea* species and *Daviesia. Doryphora sassafras* are related species are starting to colonise the damper areas of valley floor.

(vi) <u>Other Species</u>: Along the transect mosses can be found growing on the shaded areas and on the southern aspects of rocks and trees. There is evidence of seed pods of *sclerophyllous* species such as *Crowea* and *Boronia* transported by migrating and foraging birds: most of these appear to have failed to germinate successfully as evidence of young plants is absent.

Interaction between Community Member Species

Interactions which should be referred to include examples of:

(i) parasitism: e.g. mistletoe high in tree canopy,

(ii) commensalism: e.g. tree orchids living in hollows and

cracks of trees: similarly note should be made of fungal species and evidence of presence of larger bacterial species

(iii) mutualism: e.g. certain insects depend on pollen for food and these in turn contribute to and ensure successful pollination of a range of dependent species

(iv) competition: e.g. direct interference and competition between eucalyptus species for soil nutrients and access to sunlight: indirectly this competition inhibits and limits growth by weaker individuals and those living in lower levels of canopy where low levels of filtered light are common.

Food webs are helpful in that they show feeding relationships between species within the designated area. These relationships serve to highlight the distribution of energy within the system. A typical series of such relationships is presented in Figure 4.3.

An example of a study of plant communities for a recent EIS is contained in Appendix I.

The Study of Animal Communities

Land-Dwelling Species

The numbers, health and wellbeing of animals in any environment represent, or are a function themselves of, the total physical, chemical and biological environment, and their multiple, interdependent relationships. These are distributed throughout the diverse habitats or microhabitats of any area and within these they create unique, distinctive niches. Ecologists have learned that by identifying the species occupying each niche, it is possible to work out, with reasonable reliability, the populations present and their interactions.

The first step in this process is to recognise all the

different types of areas in which animals might live, and estimate the percentage of the ecosystem which they represent. Here the ecologist needs to recognise and distinguish between areas such as rocky outcrops, tree trunks, branches and leaves of trees, bodies of water, areas of grass, areas covered by leaf litter, large patches of animal faeces such as cow dung, etc. Each of these types of habitat has distinctive features. For example, the temperature in the soil or under a rock or beneath the bark of a tree will be lower than in the atmosphere. The variation in temperature will also be lower in each of these areas as they are less exposed to the direct light and heat of the sun.

Each habitat has a different exposure to the elements such as wind and rain. An area exposed to strong wind is likely to have a lower relative humidity than more sheltered areas. This exposes inhabitants to greater risk of desiccation. Sheltered areas such as those under leaf litter will have a higher relative humidity because air tends to be trapped thus lowering gaseous exchange with the atmosphere.

Each of these habitats also varies in the amount of light it receives. An area shaded by a tree, animals living in bark or beneath rocks and leaf litter rely on sensory organs other than eyes to guide their passage and search for food because of limited light. Similarly the amount and type of food ranges widely amongst these habitats. A rock supplies no water and little nutrients to animals living nearby, in contrast to the abundant and ready supply available to an organism living in the bark of a tree. These stresses and challenges influence population numbers and integrity, including their resistance to predators, disease and external physical changes.

Animal communities are best assessed by first estimating the area covered by the species (habitat area) and the numbers and age distribution of species members (ecological density). Other important parameters are: biological species classification and population size, trophic level occupied and its biomass. Habitat type, e.g. aquatic or land dwelling, whether it occupies that habitat exclusively and its capacity to also occupy other habitat-types are also required. Field researchers spend extended periods in the area under study observing animal behaviour, noting when a particular species feeds, the type of food, where the species members shelter, peer relationships within the communities and between different sexes, mating behaviour, breeding patterns, care of the young and survival. As well the ecologist maps the passage of the animal through its habitat, looking for tell-tale signs such as footprints, faeces and evidence of feeding; e.g. loss of low foliage.

This data serves as an excellent source of baseline information. Variations in numbers, behaviour and breeding often follow and accompany major changes to a habitat, as is occasioned by significant changes such as open-cut mining, increased human settlement for housing or provision of an access road into an area highly subject to fires.

Aquatic and Freshwater Ecosystems

The same principles of study are applied when studying animals and freshwater ecosystems. The methods of observation and sampling are different because humans are unable to move freely through and live in water in the same way as is possible on land. More emphasis is placed upon sampling the members of this ecosystem. This involves taking careful note of the depths in the water body samples, because temperature and the availability of nutrient can vary in water bodies owing to currents or differences in the density and physics of the water.

Most inland waterways have experienced massive increase in primary production caused by a massive increase in nutrient input from increased soil input (from rural or urban runoff), fertiliser washed in or treated sewage effluent discharged into the water. Construction of roads, dams, flood mitigation works as well as changes in drainage due to residential developments have altered flows in streams and affected their ecosystems. The introduction of exotic species such as carp, whose populations have been encouraged by prevailing conditions, has caused marked changes in the natural balance of species.

Concluding Remarks

It is clear that much technical information and a high level of expertise goes into the preparation of an EIS. In the first place, relevant environmental legislation ensures that a qualified, independent assessment of the data and its analysis is undertaken by the relevant authority responsible for approving or rejecting the project.

Assuming that the EIS data has been collected accurately and presented fairly, any decision to criticise it will need to be based on firm grounds and sourced in equally or superior and competent studies. New data of a quantitative or qualitative nature can be very helpful as it re-opens any debate through additional information. Through careful systematic work, the public participation process can and does work, but the path is not easy.

CHAPTER 5
CONSIDERING THE DATA:
What do we do with it? How do we make use of it?

What have we learned thus far? In the previous chapter, we selected plant and animal communities to introduce as examples of two field studies which are undertaken in the collection of baseline data for an EIS.

Probably most would agree now that the task of collecting and organising the data for an EIS is best left to the professionals - the scientists and technical experts trained in their different speciality areas. Putting the information together and interpreting it demands a more generalist approach and requires a cross-disciplinary scientist or a team, able to manage and balance the vociferous interests of opposing lobby-groups. As well, balancing the interests of social and community, commercial and government interests, of economic benefits against biological losses and health risks, requires a sense of vision, combined with political pragmatism and perhaps a dash of recklessness.

Considering the Data

The Public Rule

When do members of the public gain the opportunity to make comments and express their views about a proposal? Current legislation protects the interests of ordinary people, as we saw in earlier chapters (Chapters 1 and 3). Ideally, the local community likely to be affected by a development needs to know about it as early as possible in the planning process in order to ensure that misunderstanding, opposition and conflicts can be avoided. They need to learn about the proposal in a gradual, non-emotive manner.

Access to the plans and an opportunity to learn more about all aspects of a proposal are necessary to allay public concerns. Resentment, suspicion, doubt, prejudice and disharmony all flourish in an environment where secrecy exists and limits the release of information which is sought. In many instances, details are withheld for commercial reasons. Developers tend to operate in a highly competitive climate where new plant operations can result in substantial advantage over a rival.

Full details and an honest description of the risks and hazards is needed to balance the benefits of any new growth. Wide rather than limited distribution is necessary to ensure that everyone is given an opportunity to learn about the project.

Early Involvement by the Public

More and more, developers are realising that costly delays are inevitable and considerable public goodwill will be lost if residents and an affected community are not informed of intentions well before final plans are drawn.

Recent moves to invite early involvement by the public and to invite their participation make good political sense. When the developer takes the initiative, he seeks out local people, members of the affected community and tells them about his proposal, asking for comments and criticisms. The

developer can also prepare the community for scientific researchers keen to collect baseline data on the area for the
proposal EIS.

Early involvement of the public also carries a covert
benefit to the developer. It tends to lock all the participants
in the decision-making process, into the final decision -
whether or not they wholeheartedly endorse it.

The key to the success of the public involvement process, therefore, is three-fold. First, the provision of full and
easy access to all information about the project is vital. Second, there must be adequate opportunity available for concerned individuals to voice their concerns directly to the
developer and the responsible officers. Third, developers,
representatives and community agitators must all be prepared to work together and to compromise to achieve a
resolution of the issue to best serve the interests of all.

The media's tendency to sensationalise any environmental issue feeds on public scepticism of major projects
and any 'solutions' to perceived problems, when they are
fashioned by persons outside an affected community. For
this reason, it is in the interests of the proposer of a new
project, to establish communication with the local community and encourage active participation of local people in
decision-making processes. Furthermore that communication needs to be genuine and not simply a marketing exercise which can be cynically viewed as 'selling' of the idea by
a vested interest group.

Of course, keeping the public happy and involved is
just one benefit from early contact. A further benefit is
achieved from the collection of baseline data, especially in
respect of human and other biosystems as evident prior to
commencement of any proposed work.

Dealing with the Problems

How are the public concerns dealt with? Is it likely that

public opposition can actually prevent a development going ahead, or is public effort only ever likely to delay the project and result in minor "cosmetic" changes?

In most developed countries, the public is invited to express its concerns at specific times and in a proscribed manner, either by written submission or through public presentation or meeting. In the first instance, the developer usually advises the community by public announcements in the media: e.g. newspaper advertisements. Other methods include: letters to local residents deemed to be affected by the proposed development, gossip, rumours and other informal networks.

It is certainly true to observe that public comments are sought rather late in the approval process. It is equally fair to ask what value public opposition and comment could achieve at such a late date. Often significant damage to a site may have occurred when measurements were made, initial surveys drawn and preliminary investigatory work undertaken in drawing up the necessary documentation for project approval.

Nonetheless, and despite the apparent lateness of any public involvement, projects can still be de-railed at this seemingly late stage. The key to success is that concerns, when expressed, are genuine: that they are backed by substantive and well researched data, and that they are well publicised and correctly presented.

When a public submission is dealt with, it is customary for each issue raised to be addressed, in turn. Costly modifications, greater occupational and health controls and even compensation mechanisms can be forced into the original proposals.

During consideration of plans for Sydney's Third Airport, three separate EISs were sought as a result of public opposition to the original, and then amended proposals. Each new submission demanded changes to meet public concerns. The project was finally approved and the first sod of soil turned more than four years later than the original planned date for commencement of the project.

Some projects have failed to go ahead because of the weight of public protest. Others fall by the wayside when financiers withdraw their support because of the prohibitive costs associated with special environmental mitigation measures, clean-up operations, and health and safety provisions sought by conservationists and health experts. Examples include the cancellation of a large woodpulp and newsprint mill at Wesley Vale in northern Tasmania due to the stringency of environmental controls, and of a gold mining at Coronation Hill in the boundaries of Kakadu Park in the Northern Territory because of concerns relating to aboriginal 'dreamings' about a mythical figure called Bula.

In most instances, however, projects do finally go ahead - subject to major modification. These mitigating measures range from the inclusion of filters to reduce pollutant emissions, to screen plantings to conceal unsightly plant and buildings, to restoration measures such as landscaped gardens, planting of forests and re-vegetation.

Relentless, unrelieved and constant public pressure backed by accurate scientific data is often necessary to ensure that mitigating measures address the problems which the affected public view as worst. Frequently, public priorities and concerns differ substantially from those of the developer. Consultants employed by the developer often express amazement following meetings with representatives of local communities. Whereas consultants may have expected local people to be most concerned about health risks posed by pollutant emissions, their representatives may rank noise and visual landscape changes as prior concerns.

On many occasions, mechanisms may be required to compensate parties adversely affected by a development, as well as extensive mitigating measures. Under these circumstances, claims need to be based on sound scientific grounds and not take the form of simplistic, highly emotional idealism. These would be doomed to inevitable failure. Any evidence must be collected and assessed by qualified researchers. Quantitative collection methods exist for technical and physical data. Qualitative measurements of such sectors as visual effects of a development, changes in ethnic population mix, loss of amenities, traffic hazards and job opportunities are less easy to record. Public sentiment, in particular, is especially hard to monitor and is highly changeable because it often reflects other broader political views and community beliefs.

Public dissent and resistance to a proposal will normally result in an aggressive defence from the developer. However, the larger the project, the more wealthy the developer, the harder and the longer the campaign needs to be. A prime example is the World Bank funded Namada Dam project in northern India.

By the time public submissions are sought, it is highly likely that the developer may have already begun clearance of the site and constructed temporary buildings, foundations, etc. Yet these 'minor' changes do alter the environment, regardless of how imperceptible they may appear on the surface. Local councils and government authorities keen to encourage growth and development in their areas can give tacit but unofficial approval, allowing 'minor' intrusions such as access roads and huts with the intention of aiding in the data collection process.

The Rational Consideration of a Development Application

The public and planners alike expect any development to be accused of directly or indirectly harming the natural environment. There are many direct and indirect benefits, however, aside from increased employment, higher wages, job-opportunities for school leavers, improved public facilities and transport. Lost open space can be replaced by better amenities, including social facilities for recreation which were lacking. New wildlife conservation areas can be set aside and new species identified and classified as a result of discovery.

In most western and developed nations, laws have been enacted which specify the need for environmental impact assessment when different types of projects are planned and developments are likely to significantly alter an existing environment. In a previous Chapter, Table 2.2 summarised those projects where mandatory EISs were usually required. Where the process is well developed, there are detailed guidelines which define the level of detail required and the range of impacts on which information is required.

Table 5.1 focuses on the criteria used to determine the need for EIA. Areas where public consultation and review are possible is further clarified in Figure 5.1.

Table 5.1 . Minimum data set for industrial development proposals
Assessment of short and long-term effects on all living species, including human beings.

Information on environmental, social and economic factors which are likely to influence the susceptibility of a subject human population.

Information on climatic, geological and ecological factors which are likely to influence the susceptibility of plants and animals in the affected area.

Inventory of pollutants and concentrations likely to be released into the environment.

Quantitative description of diseases related to pollutants associated with the development.

Quantitative description of the dispersive, absorptive and adsorptive mechanisms in air, water and soil.

Considering the Data

Prediction of concentrations of the pollutants over the various periods of time to which the affected populations will be exposed.

Forecasting of health effects, including their intensity, duration and lag time between exposure and effect.

Identification of the means by which effects can be prevented, minimised or eliminated, together with relevant costs.

Assessment of impacts of alternatives.

Definition of the monitoring systems to be instituted together with the activities of the proposed project.

'Thinking Globally, Acting Locally'

Local people can take up the challenge of effecting change in project planning in a number ways. Resident action groups have proved extremely effective avenues for communication within a neighbourhood. Members can actively engage in the collection and distribution of information. They usually trigger education on the project within the local community; e.g. by organising local meetings and discussions groups, and also formal forums with project planners. Resident action groups can also lobby the responsible planning bodies for changes to decisions, for strengthening of standards and screening requirements, and for the inclusion of strict monitoring procedures in any final submission.

Among the features shared by local action groups are:
* they form and dissolve around a specific proposal or initiative, and thus tend to have a temporary status
* usually formed on a voluntary basis around a small group of committed, highly motivated individuals
* provide excellent source for information on specific project concerns: sometimes characterised by highly professional and competent research of scientific data
* rapidly gain skills in media and political liaison

Environmental Impact Assessment

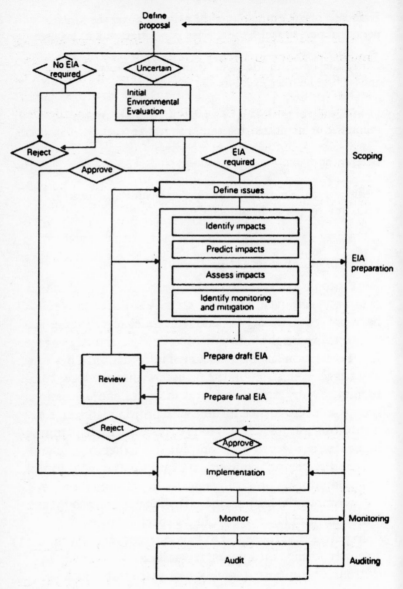

Flow diagram showing the main elements of an EIA process.

How To Achieve Local Participation

A useful starting point for the exploration of just how one might go about involving people in a complex decision-making process has been recently provided in Australia. As a practical way of reaching decisions on a matter where the community had expressed strong feelings, the Independent Panel on Intractable Waste, refered to earlier, was established by joint agreement between Australian, NSW and Victorian government ministers to review and resolve the increasing accumulation of intractable wastes. The panel defined four objectives for community consultation:

(i) the appraisal of the range and intensity of community opinion

(ii) the invitation to bring forward any relevant technical and environmental information

(iii) the education of the community through the consultative process

(iv) the design of a strategy that is broadly acceptable to the community.

In this exercise, the Panel defined the community as 'government, industry, research and scientific bodies, representative or sectional organisations and individual citizens.'

In order to harness the required knowledge and opinion, the Independent Panel adopted a three-step approach involving:

* the determination of overall community understanding and attitudes

* the meeting with selected groups which need to be party to a satisfactory resolution

* the final negotiation of a report prepared for the three government bodies that initially established the Panel.

The underlying belief that formed the basis for the consultative approach was that the process needed to be 'pragmatic, flexible and iterative rather than preordained and highly structured'. A successful outcome was seen to require a capacity to manage the ambiguity of the issue not to engineer a predetermined solution.

The importance of this particular consultative approach arises from the fact that it has the basics for success securely in place. First, there is no predetermined solution. Second, there is commitment to the maintenance of an open mind and all parties are seen to benefit from the process of on-going education. The eventual action that will be undertaken will not be due to the power or authority of any one sectional interest, but will represent an outcome that is broadly acceptable to the community.

In essence, to achieve meaningful local participation and subsequently cooperation of other stakeholders, it needs to be accepted that there are different but none the less equally valid sorts of knowledge. These include the domains of historical knowledge (the domain of the locals), scientific knowledge (possessed by scientists), emotional knowledge (for the committed), technical knowledge (technocrats) and communication knowledge. In this way all involved become co-researchers and thus share in the ownership of the eventual outcome. Having accepted the ownership of the resolution, it is then the ethical response to accept the consequences that flow from the action, including the often unexpected and unwanted side effects.

The Community in the Monitoring Role

Thus far, no effort has been made to detail legislation and talk about the use of sanctions and penalties to limit pollution and to enforce compliance. Where there is public concern about a proposal and processes used, those responsible for framing public concerns are advised by the legally trained on the exact sections of legislation involved, especially its limitations.

Individuals often feel they have evidence of breaches of pollution, and/or a valid claim for compensation e.g. decline in land values due to siting of an unpopular facility near their home, or health costs arising as a result of exposure to toxic chemicals in the course of employment. Public interest bodies and community resource centres can advise on proper procedures to initiate a claim, and to obtain appropriate professional assistance from sympathetic parties who possess the expertise to argue before a judicial body.

The exercise of monitoring and policing in itself is based on the belief that the community can be trusted to accurately collect and record technical data as unpaid field observers. The NSW Water Board conducts a highly successful water quality monitoring scheme involving children from over 80 schools across the state. Called Streamwatch, the scheme requires students, under supervision from their science teachers, regularly to collect samples and transmit to a central authority, data on turbidity, pH, faecal bacterial count, smell and taste from important local rivers and streams. Professional officers follow up on instances where high levels of phosphates, heavy metals or anaerobic organisms are detected, so that appropriate warnings can be issued to affected communities.

Contingency measures are needed to facilitate detection and timely response to potential problems and breaches. The public can assist in the drawing up of licencing requirements - inconsistencies do occur and innocent householder actions for personal use do not come within the ambit of the law. Emergency provisions and procedures to deal with accidents involving acute exposure to a noxious and toxic agency are matters for government authorities. Members of the public frequently comment on the adequacy or otherwise of these provisions. Community pressure is often necessary to ensure the preparation and maintenance of an active warning and alert system.

Mitigation Practice, Monitoring and Human Health Impacts

Finally, it is clear from the preceding discussion that the

public and individuals have a very real role to play in monitoring environmental impacts, particularly prior to the commencement of operations. At this point, the accurate and careful collection of baseline data is most important. The anticipated effects serve as a basis for the planning of mitigating measures to minimise likely impacts. Compensatory mechanisms can also be established to redress proven adverse impacts. Suggestions of the likely measures are contained in the following Table 5.2.

Clearly this table merely summarises but some of the likely health effects and possible mitigating measures which might be raised in the environmental impact statement associated with a proposal. Each project is different and acceptable measures range from those of low cost to others involving vast sums for dubious value.

In the final analysis, where high cost measures are demanded by the planning authority, large development projects may be cancelled or permanently delayed resulting in substantial loss of opportunity. This, of course, is the ultimate and final answer for the developer frustrated by continual opposition and hampered by restrictive laws and stringent regulations. The Wesley Vale pulp and paper mill is one such example. Sadly, these results often follow continued harassment from a committed minority in a community, whose members are otherwise well-inclined to the proposed project. Herein lies the ultimate tragedy and failure of the EIA legislation for all who might otherwise have gained from it.

Table 5.2. Environmental health factors and related disease agents, exposure, risk groups and mitigation measures

Air pollution
Factors and their effects:
inert materials and dusts (irritation of respiratory tract)
pathogens on aerosols (respiratory diseases)
gaseous or suspended particulate toxic chemicals (carcinogens)
 oxygen deficit (asphyxia)
Exposure:
people breathe indoor and urban air
Risk groups:
people with chronic respiratory diseases
Mitigation:
abatement of emissions at source
dispersion of pollutants in higher atmosphere
reduction in exposure of risk groups

Solid wastes improperly disposed
Factors and Their Effects:
inert materials such as stone, glass and metal (injury hazard)
toxic materials (human ingestion through water or food)
organic fermentation products (favours growth of pathogens)
food residues (increase population of disease animal vectors)
Exposure:
contact with disease vectors
contact with toxic materials
consuming contaminated food or water
Risk Groups:
children playing on discharge sites
garbage collection workers
consumers of water from aquifers contaminated by leachate
people within dispersal range of vectors
Mitigation Measures:
proper selection of disposal sites
fencing of disposal sites
burying disposed waste under soil cover

Source: after Giroult, in Wathern *et al*, 1988, p.267

CHAPTER 6:
PUTTING IT ALL TOGETHER

Our introduction to environmental impact assessment processes is drawing to a close. We have seen that it involves two basic steps:
1. preparation of an environmental impact statement (EIS) by the proponent of a project, and
2. review of the EIS by the public and government officers to evaluate its accuracy and obtain comment as to whether or not the project should proceed.
   Does the exercise work? And for whom does it work?

The Effectiveness of the EIA Process
The effectiveness of an environmental impact study depends upon the study or report being:
* comprehensive
* readily available to the public and with sufficient time allowed for the preparation of submissions
* an evaluation by a fair and impartial tribunal
* subject to rejection by the tribunal if it was improperly withheld, not comprehensive or was substantially inaccurate.

  Additional safeguards needed include:
* adjudication through a public inquiry in a major or controversial issue

* a review of disputed matters in a court of law
* a mechanism for monitoring compliance with the terms of the project approval.

It has already been observed that environmental legislation is of varying quality. It also varies between states, even within the one country. Most parties agree, however, that the inherent fault with the whole EIA process in most developed countries is that the environmental impact statement is the product of work undertaken by consultants hired by the proponents.

Realistically therefore, the system only really works if all major studies are tested in a public environment inquiry under independent judges. Bearing in mind the number of EISs produced, this is necessarily unwieldy. It would be extremely costly and greatly increase time delays.

One of the major criticisms coming from developers who have had experience with the EIA process is that too many are adjudicated on by the proponent government department. Some of the most experienced environmental groups have also grown cynical of the apparent wastage of their time in submitting responses. Smaller public and community groups are at an even greater disadvantage.

These criticisms suggest that if the adjudication process cannot be taken from the hands of proponent departments and placed in those of a truly independent body, then the mechanism is doomed. For the present, it seems that community groups can achieve better results through practical lobbying, media skills and aggressive public campaigns.

Winners and Losers

Who benefits from this process? Superficially it would seem that the party least likely to benefit from EIA would be the project proponents. It must be remembered, however, that proponents who integrate the EIA process into their planning can expect a more publicly acceptable project. The exploration of the project alternatives is usually done in the

EIS. In some cases, this can actually result in an improved or even less costly project design. Maximum benefit is achieved if alternative proposals are integrated into the planning stages of a proposal.

Another criticism often directed against EIA is that the process is used to justify decisions which have already been made. This gripe is voiced when the review does not lead to any modification of the original proposal, even where a lengthy public consultation has been followed and objections appear to have been ignored. In this instance, critics of the project would argue that the public consultation stage was merely a formality and that their time and efforts preparing submissions were wasted. These same groups then excuse activities such as picketing and demonstrating, claiming that these methods are the only effective means to ensure that their concerns will be resolved as judged appropriate.

Successes and Failures of the EIA Process

Today, some years after its inception, the EIA process can list notable successes and spectacular failures. The following examples serve to substantiate the arguments and principles discussed here. Because Australia is among those nations at the forefront of large-scale commercial development proposals and has a particularly fragile environment, parts of which are subject to claims by indigenous peoples, the examples quoted are drawn from there. However, the same principles are equally applicable to more developed environments such as Britain and Europe. In these countries, the legislative and public advocacy focus will be upon site reclamation and clean-up requirements which must preceed new development and re-zoning.

The answer to any query regarding the success or failure of a project is likely to depend on the individuals approached, what part they played in opposing or promoting the project, and whether they still hold the same views now that the development has been completed. To an extent, this will depend on their own personal involvement in the project, and whether they consider that they gained or lost in the negotiations.

Two significant successes from negotiations arising in the EIA process in Australia have been the Tunnel under Sydney Harbour (a project similar to the Chunnel connecting France with Britain), and the Franklin River Hydroelectric Scheme proposed in the 1970s for south-western Tasmania. At the time, this last issue attracted worldwide publicity due to involvement in protests by the British conservationist David Bellamy.

The Sydney Tunnel was completed in 1992, and results thus far, have confirmed predicted impacts of the project on the environment and benefits to commuters to the city. Substantial and irreparable damage anticipated by opponents was rejected during consideration of the EIS. Compromises were made where warranted and dire consequences have not been forthcoming.

In the case of the Franklin and Gordon Rivers, limited development for hydroelectric power generation has meant that large areas of land have been declared as heritage reserves. Wildlife, native birds and plants have successfully recolonised the original project area. Contrary to predictions from environmentalists, species diversity has also increased with the flooding of Lake Pedder which serves as catchment for the river system, and negative impacts on the surrounding environment have been minimal.

Thus, as a result of public controversy surrounding both original proposals, many modifications were sought and final decisions involved compromises from all parties.

Failure of the environmental impact assessment process also depends on that failure being defined. The preparation of an EIS is a mechanism to proponents of projects to consider the effects of their activities, on others, and over a long period of time. The EIA process serves as a check on hasty action and thereby provides the opportunity for external review and evaluation of any project. An example of such a check on fast-tracking of developments was evident in a review of Australian government action to speed development of the Mt. Todd mine in northern Australia in late 1992.

The need to prepare an EIS, therefore, and the fact that it will be critically examined by outside agencies, gives project opponents greater leverage in successfully opposing new developments. It could be argued that the proposed Coronation Hill Gold Mining venture referred to earlier in northern Australia was abandoned because community interests and powerful political lobby-groups were able to mount a campaign which forced a major change in government policy. For the developers involved, the rejection of the proposed gold mine at Coronation Hill within the boundaries of the Kakadu National Park in Northern Territory, Australia can be cited as evidence of such action. After considerable preliminary negotiation, the preparation of a detailed and expensive environmental impact study and assurances of support from the Australian government, the approval to mine was denied to the developer consortium. As has been referred to earlier, the decision was based on both conservation grounds and claims of infringement of the rights of the Jawoyn people, the traditional aboriginal owners of the land.

These ventures illustrate the high level of uncertainty associated with major projects in countries where governments impose sophisticated and detailed controls including EIS requirements. It could also be said to serve as a warning to future project venturers on the need for a clearer definition of resources management policy to be applied to an area by government, prior to any attempt to develop any portion of that area.

The conservation movement regarded the rejection of mining at Coronation Hill mine as a victory for the conservation movement. How did this occur? Politics, fuelled by effective lobbying by conservation groups changed public opinion. Where no defined government policy could be cited to apply to the project, the objections of the affected community gained precedence in the eyes of the courts. The actions of government, the failure of existing policy to encompass the proposal, and the deficiencies in the law exposed were roundly condemned by the business community.

Effectively, however, the Coronation Hill decision represented a major shift in government policy towards development projects and subsequently resulted in the cancellation or shelving of a number of major mining proposals under consideration in Australia. To this day, the project proposal for Coronation Hill and other major projects in Australia remain merely that. They await a change in government or mechanisms to fast-track such proposals in a regularised manner where future benefits to the nation warrant such actions.

Clearing Up Misunderstandings

A misconception commonly held by members of the public and also in some sections of government, is that the strong objections voiced by developers to the approvals procedure stem from requirements to incorporate environmental protection and management into the design and operation of new projects. On the contrary, the need for and value of such measures are now widely accepted by the industry. The real concern of developers arises from the uncertainty and the degree of financial risk to which they are exposed as a result of the administrative delays associated with the approvals process. This may be up to 3-5 years, whereas 12-18 months would appear more realistic.

The other major concern is the cost implications of the procedures. These can range from $US50,000 to $US1 million simply for direct EIS costs alone, with consequent costs possibly far greater. Further difficulties for the developer arise because of the constraints and inflexibility of the system, the dynamism or constant changes occurring to requirements, and the uncertainties and differences between and within government departments.

The greatest source of uncertainty is probably the overlap in the approvals procedure itself. A senior executive of a major coal exporter in Australia recently observed that to obtain approval for a major coal mine in that country may involve up to 29 departments at all levels of government, and require 25 'approvals' from twelve institutional sources. The uncertainty and time delays are thus compounded by the variability of the demands of the various departments.

Experience has turned these many problems into 'facts of life' for the developer. The result is that there must be very real gains offering for a complex proposal to go ahead. It also means that smaller players with lesser resources tend to be 'frozen out', leaving all but the largest projects. These are usually highly syndicated projects involving international consultancy teams, substantial capital investment and complex organisation.

A Word on Ethics

Engineers, many scientists and health professionals belong to professional associations and, as members, are obliged to comply with standards and ethics to ensure that the reputation of the profession as a whole is preserved. The same engineers, scientists and personnel involved in putting together environmental impact statements are often faced with ethical dilemmas. The nexus of these dilemmas is the conflict between duty to an employer or client and responsibility to the health and welfare of the wider community.

The first problem lies in the preparatory phase of the project data. Environmental impact statements (or studies) are prepared or commissioned by proponents for a development. The careers and promotional prospects of the involved engineers and scientists are mortgaged to their employer, whose prime objective is to gain approval for the project.

It is to be assumed that in most instances these professionals try to act with honesty, integrity and objectivity at all times. With these pressures, it is not overly cynical to expect that people reading the EIS might bear in mind that it is highly likely that there has been a favourable interpretation of data. On occasions, engineers responsible for technical data have observed that final reports have omitted crucial figures. Opinions alone are of no substantive value unless supported by facts. In addition, it sometimes happens that emphasis is given to the more favourable aspects and only passing reference to adverse features. Thus the coordinator of the submission may find himself responsible for 'gilding the lily'. Imprecise language and the adoption of a summary type format can also mask facts and give a misleading impression.

Final Remarks

Well here we are at the end. Now is the time to wish you luck. You may be involved in the preparation phase of the EIS itself, as a technical or non-technical member of the team and anxious to know more about how it all ties together. You may be unhappy about a project proposed for your suburb and unable to understand the documents you have been invited to view, as a member of the local community. You may even be preparing a public submission to an EIS because you are strongly opposed to a new development and feel it important that opposition be expressed.

The previous six chapters have outlined your task. Follow the steps, seek professional and technical help as needed from friends, government agencies and even the proponent of the project itself. When you have put it all together, run it past someone who has had experience with the EIA process, for example a public advocacy group or an environmental group, for their opinion about how you have framed your arguments.

It is highly likely that if you have not had much experience with the political process, you may need to substantially re-organise and rephrase your arguments. Do not be dissuaded. Press on and when ready, present it, clean, neat - and before the due date and time. Good Luck!

Appendix
Field Study of a Plant Community

*PLANT COMMUNITIES ON HAWKESBURY
SANDSTONE IN THE UPPER LANE COVE
VALLEY. KISSING POINT ROAD SOUTH
TURRAMURRA.*

Author:
Commissioned by:
Date of Study: October - November 1992.
Date of Release of Report: 30 November 1992.

Table of Contents

1. Introduction

> Sydney sandstone.
> Twisted orange bark of the red gum
> on dissected heights.
> Soft cream and green of flannel flowers,
> beneath serrated leaves of banksia.
> Brightness of white cockatoo,
> with sunclad crest in the turpentines.

1.1 South Turramurra Bushland is part of one of two valleys, Middle Harbour Creek and the Lane Cove River, deeply incised into the Hawkesbury Sandstone in the northern part of Sydney.

South Turramurra Bushland is shown on maps 173 and 174 in the UBD Sydney 1992.

It is located on the eastern side of the Upper Lane Cove River.

2. Structure

The structure of the community intersected by the transect is open forest dominated by *Eucalyptus piperita* (Sydney Peppermint) - *Angophora costata* (Sydney Red Gum) association. Tree height is a maximum of 15.5 meters with canopy height greater than bole height. *Banksia Serrata* (Old Man Banksia) forms a second layer of large shrubs / small trees between 2 and 7 meters. This incomplete tree cover (30 % - 70 %) permits entry of sunlight and ensures a well developed shrub layer with a discontinuous herb strata.

Appendix: Field Study of a Plant Community

3. Profile

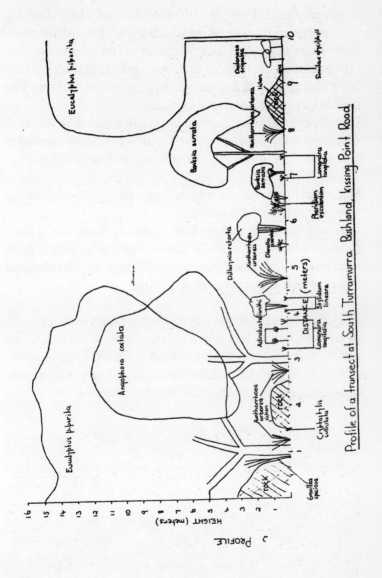

Profile of a transect at South Turramurra Bushland, Kissing Point Road.

4. Plant Life Forms and Features

Sclerophyllous Trees which demonstrate the following xeromorphic characteristics which allow these plants to reduce water loss, especially from leaves and stems:
- deep, extensive root systems. Angophora costata particularly shows this characteristic with its roots flowing like treacle over rocks and into crevices.
- small cells with thick, often lignefied walls. The leaves of eucalypts are hard and leathery which restrict evaporation and water loss at the major site of water loss in the plant.

Sclerophyllous Shrubs which demonstrate the following xeromorphic characteristics:
- factors limiting surface transpiration from the leaves. *Banksia serrata* particularly shows this characteristic with thickened cuticle and supportive cells inside its sclerphyllous leaves which allow it to resist wilting.

Rock outcrops shown on the transect are Hawkesbury Sandstone which was laid down in the Triassic Period. This parent rock weathers slowly to a nutritionally impoverished, quartz rich, skeletal soil, which is low in organic matter, deficient in nitrogen and poor in phosphorous.

Pteridium esculentum (Common Bracken)
If the sclerophyllous bush is burned too frequently there is a tendency for it to be replaced by this very common fern which can form dense stands. Otherwise it grows in a wide variety of habitats, on the shady edges of sand dunes and from the coast to the tablelands in dry forest on well drained soils. The fronds rise to 3 m. straight up from a long - creeping underground stem which is clothed in reddish hairs. In this study this patch of bracken has been burned by an arsonist. It is a small area within the study area.

Appendix: Field Study of a Plant Community

Lomandra longifolia and *Dianella caerulea:* are herbaceous plants which occupy shallow pockets of sandy loam.

5. Interactions

The surprising profusion of plants which characterise the open forests on Hawkesbury Sandstone is a reaction to the infertility of the soil. Each family contributes to the efficient use or production of nutrients.

Eucalyptus piperita (Sydney Peppermint) - Angophora costata (Sydney Red Gum)

This open forest association is part of what is popularly regarded as Sydney's 'bush'. It is certainly part of the essence of what it feels to be Australian. This is generated by the genetic make up of eucalypts which determines their shape, development, flowers and fruit, distinctive scent and the unmistakable way in which they grow.

Insect predation is an important part of this association, and many wood and foliage feeding insects and their host plants have a long history of coevolution. Insects such as termites, beetles, phasmids, caterpillars, sawfly larvae eat 15 % - 20 % of the surface area of eucalypt leaves every year. This is a limiting factor as the insects act as regulators of the rate at which solar energy is fixed by photosynthesis for the primary production of the forest ecosystem.

However insect predators also have important consequences on mineral cycling, plant structure and diversity and energy distribution. Severe defoliation obviously affect the physiological status of the host; weakening its competitive ability by limiting its growth. But leaf eating also causes increased litter fall of insect excrement and bodies, leaves and leaf parts and branches that add nutrients to the soil /

litter systems. This litter fall also contributes to the soil structure. Weakened, old and suppressed plants often die after insect defoliation; thus adding to the soil / litter nutrient balance and providing more nutrients, light, heat and moisture for the remaining plants.

Insect predators may help to maintain nutrient cycling and primary production at the most optimal rate for a particular site and system by facilitating the distribution of energy and circulation of matter.

The ability to recover from defoliation is a prime requirement for eucalyptus which live in environments where fire, termites, fungi, insect predators and drought are common features.

There are four ways in which eucalyptus can produce new growth in the form of leafy shoots. The primary means by which these new shoots are developed is from naked buds which is the method that gives a tree almost unlimited growth potential.

The other three (accessory buds, epicormic shoots and lignotuber) are reserves; and if the growing tip is destroyed these reserves give the trees their extraordinary hardiness and persistence. Poor soils, bushfires, defoliation by insects, trunks broken by storms; all act as regulators on the growth and vitality of eucalypts.

Banksia serrata (Old Man Banksia) has a mutual association with mycorrhyzal fungi. This association occurs especially on infertile soils where it is an efficient collector of phosphates. Therefore it is particularly important in woodlands and sclerophyll forests.

The fungi become intimately associated with the roots of

the plant. Sometimes the fungal hyphae penetrate the root tissues. These are better able to absorb minerals from infertile soils than the plant roots and they pass these minerals onto the plant. The plant passes on sugars and other growth substances to the fungi.

The mycorrhyzal fungi may be involved in the following in its association with Banksia serrata:
- promoting rapid transport of water to the plant than through the soil.
- counteracting conditions which are deleterious to the plant roots such as high acidity, high aluminium levels, high soil temperatures, and root disease.
- improving soil structure.
- enhancing nutrient conservation.
- protection of the partner from pathogenic fungi and bacteria.

Dillwynia retorta (Eggs and Bacon) is a shrub to 3 m. The feature that distinguishes it is the fine leaf, 4 - 12 mm long, which has a conspicuous twist.

It is one of a number of species (Fabaceae, Allocasuani, Acacia) which live in close symbiotic relationship with the nitrogen fixing bacterium Rhizobium. Nitrates are provided by the 'fixing' of nitrogen in the root nodules with the help of the Rhizobium. It is the first to grow after a nitrate - depleting fire.

Cryptostylis subulata (Large Tongue Orchid) This evergreen terrestrial orchid often grows in large colonies. Its preferred habitat is damp sandy soils in open forest and heath. In late spring a terminal raceme of 2 - 14 green to red flowers rises from the radical, lance-shaped leaves 5 - 10 cm long.

It is also one of a number of plants which use mimicry as

part of pollination. Orchid flowers are in fact the most impressive mimics. Such flowers look so much like particular female insects that males of that species attempt to copulate with them. During pseudo-copulation and subsequent visits to other orchids pollination may be effected by the male wasp. The genus *Cryptostylis* is pollinated by males of Lissopimpla semipunctata (a species of ichneumon wasp).

Xanthorrhoea arborea (Grasstree) is often seen growing with *Angophora costata* on the rocky hillsides of Sydney and epitomises the 'bush'. The resin impregnated trunk varies from scarcely apparent to 2 m tall, flourishing a tuft of long grass-like leaves up to 1.5 m long. Cream flowers massed in a complex cylindrical spike on a pole to 2m. Birds and insects, particularly butterflies, flock to its heavy nectar flow in spring. The leaf bases of grasstrees are rich in a non-flammable insulating resin which protects the plant from fire. The rest of the leaf burns readily. These plants shoot very rapidly after fire.

Grevillea speciosa (Red Spider Flower) formally known as *Grevillea punicea*. The colour (scarlet) of the flower is stunning and in spring this either upright or bushy shrub is very colourful. It grows to about 2.5 m., with 5 - I 0 cm ovate to lance-shaped leaves. It produces copious quantities of nectar pollen. An important secondary source of food for birds visiting these flowers is the large numbers of insects they also attract. These are robber organisms which feed on nectar and pollen produced without causing pollination.

Grevillea speciosa is one of many bird pollinated flowers and shows coevolution between the flower-visiting animals and a particular group of floral characteristics.

The recurved style of grevillea brings its stigma in contact with stamens where it collects pollen. The style then straightens and waits for a pollinator. The nectar is at the base of the flower at this stage. An eastern spinebill drinks nectar and pollen is brushed onto its head. When the pollen is all removed or decayed the stigma becomes receptive to pollen and the flower waits for another spinebill or other honeyeater with pollen on its head to pollinate it. Therefore the flower must be visited twice to be pollinated.

The family *Myrtaceae* represented in this study by *Kunzea ambigua* (Butterfly Bush) and *Leptospermum attenuatum* are keystone mutuals in open forest, woodlands and heath because of the timing of their flowering, significant quantities of nectar and pollen they produce, and the large number of species which feed on their nectar and pollen at times when not much food is available elsewhere.

Pteridium esculentum (Common Bracken) If the sclerophyllous bush is burned too frequently there is a tendency for it to be replaced by this very common fern which can form dense stands. Otherwise it grows in a wide variety of habitats, on the shady edges of sand dunes and from the coast to the tablelands in dry forest on well drained soils. The fronds rise to 3 m straight up from a long - creeping underground stem which is clothed in reddish hairs. In this study this area of bracken has been burned, probably by an arsonist. It is a small area with in the study area.

Ants are a keystone species in communities where they patrol and defend a variety of plants such as eucalypts and wattles from predators. They regularly disperse the seeds of favoured species; boronias, wattle and peas. There is also coevolution with ants and plants with elaisomes on their seeds, as well as special glands on their stems or leaves to attract ants onto the plant.

Termites. In Hawkesbury Sandstone areas where water

stress is a frequent occurrence the activities of termites as decomposers of litter become an important source of nutrients when microbial decomposers in the soil and litter begin to shut down their activities in response to drying out. Termites contribute a great deal to the decomposition of litter, recycling of nutrients, ageing and destruction of trees and enhancement of growth of remaining trees and the provision of essential nesting hollows for lots of organisms (kingfishers, kookaburras).

Tunnels and galleries provide large pores and channels in the soil that facilitate penetration of roots and percolation of air and water into the soil.

Hawkesbury Sandstone soils become quite water repellent during droughts and dry spells. The activities of ants and termites means that water from showers which would mostly runoff and evaporate, is directed underground to be absorbed by roots.

6. Foodwebs

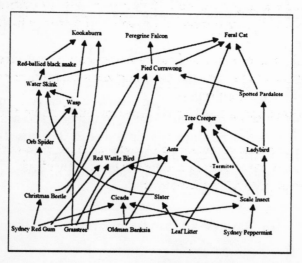

7. *Fire Regime*

In 1976 the entire urban bushland area in the Upper Lane Cove Valley was burned by an arsonist. It was a hot summer wildfire. Because this area is surrounded by housing, small fires often happen as a result of barbecues, vandals or deliberate arson. Ironically, however there is no natural fire regime for this area which is proclaimed 'urban bushland'. The area has been under spasmodic human habitation for almost two hundred years and is hence judged no longer in a pristine state.

In early 1994 another fire - also attributed to arson - again devastated the area, including a number of surrounding homes. Blame and accusation was heaped on 'greenies' and conservationists who had convinced the local Ku-ring-gai Municipal Council to reduce the frequency of Hazard Reduction Burning. This is a routine practice in bush area management where danger exists to communities and wildlife as a result of accidental and deliberate ignition of bush.

Therefore Hazard Reduction Burning takes place in high priority areas at approximately 10 year intervals. Priorities are determined for the areas by a number of factors which include:
- Fuel load, as determined by field measurements, years since and intensity of the previous burn and accumulation rates of different vegetation types.
- Landform (slope, aspect), wind direction and known fire paths.
- Size of the parcel of bushland (which can have considerable influence on the speed and intensity of a bushfire).
- Number of properties threatened.

9. Pets

Uncontrolled pets (cats and dogs) cause significant impact on this site, such as:
- killing and disturbing native fauna, especially cats.
- changing the nutrient regime, by depositing droppings to favour weed growth.
- depositing droppings in the bush, especially dogs.

8. Weeds

At least 50% of Sydney's remaining bushland is under long-term threat from weeds.

Common weeds found on this site include:
- understorey weeds (privet, asparagus fern).
- weeds of disturbed areas (roadsides, sewer lines), (blackberry, croton weed, wild strawberry, mist flower, kikuyu, lantana, plantain).

Weeds grow in disturbed areas. The primary factors causing weed growth are:
- Urbanisation which causes a change in the moisture regime which is favourable for weed species. Urbanisation reduces the normal water stress in bushland by increasing the proportion of run off after storms.
- Increased nutrients from storm water, sewage overflows, dumped garden refuse and introduced soils and stone.
- Increased light caused by disturbance which provides conditions for primary stabilising weed species.
- Weed seeds introduced mainly through storm water. Humans, animals, cats, birds and wind are also vectors.

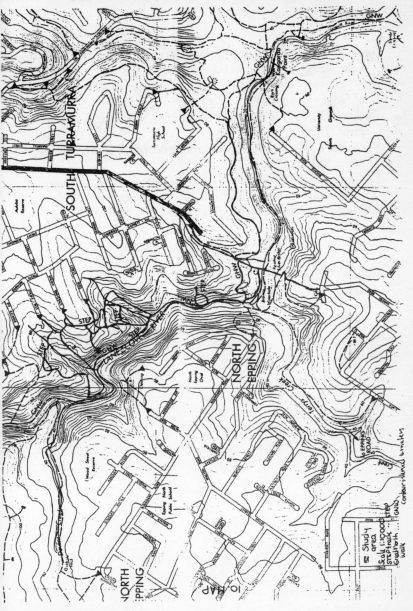

11. Bibliography

Baker Margaret, Corringhan Robin, Dark Jill. Native Plants of the Sidney Region. Three Sisters Publication. 1989.

Benson Doug, Howell Jocelyn. Taken for Granted - The Bushland of Sidney and its Suburbs. Kangaroo Press. 1990.

Buchanan Robin. Common Weeds of Sydney Bushland. Inkata Press. 1981.

Buchanan Robin. South Turramurra Bushland: A Plan of Management. STEP. 1985.

Edmonds Tony, Webb Joan. Sydney Sandstone Flora. A beginner's Guide to Native Plants. NSW University Press. 1986.

Ku-ring-gai Urban Bushland. SEPP # 19 Draft Plan of Management. July 1988.

Vandenbeld John. Nature of Australia. A portrait of the Island Continent. Collins Australia and ABC Enterprises. 1988.

Walking Tracks in the Upper Lane Cove Valley. STEP Inc. Community Based Environmental Conservation.

References/Bibliography

Australian and New Zealand Environment Council (ANZEC) 1991, 'National Principles of EIA in Australia' (Draft) Paper prepared by ANZEC for Integration in the Proposed Intergovernmental agreement on the Environment.

Australian Environment Council 1986, 'Guide to Environmental Legislation and Administrative Arrangements in Australia'. Report no. 18, AGPS, Canberra.

Australian International Development Assistance Bureau (AIDAB), 1990, 'Ecologically Sustainable Development in International Development Cooperation: an Interim Policy Statement'. Commonwealth of Australia, Canberra.

Australian International Development Assistance Bureau (AIDAB), 1991, *A Handbook for Environmental Audit.* Sector Report no. 1.

Bailey J. & Hobbs V., 1990: 'A Proposed Framework and Database for EIA Auditing', *Journal of Environmental Management,* 31: 163-172.

Bisset R. & Tomlinson P. 1988: 'Monitoring and Auditing of Impacts', in P.Wather (ed) *Environmental Impact Assessment: Theory and Practice.* Unwin Hyman, London, 117-128.

British Airways: Annual Environmental Report, September 1992.

Commonwealth of Australia, 1988, Environmental Protection (Impact of Proposals) Act, 1974.

Commonwealth of Australia, 1991, Position Paper on Proposed Commonwealth Environmental Protection Agency. AGPS,

Canberra.

de Konig H.W. 1987, *Setting Environmental Standards, Guidelines for Decision-Making*. WHO, Geneva.

Ewan C., Young A., Bryant E. & Calvert D. 1992, 'National Framework for Health Impact Assessment' in *Environmental Impact Assessment* Volumes 1 and 2. National Better Health Program, University of Wollongong.

Pearce, Fred. 'Corporate Shades of Green' in *New Scientist*, 3 October 1992, pp.21-2.

Porter C.F., 1985, *Environmental Impact Assessment: A Practical Guide*. University of Queensland Press, St.Lucia.

Rogers J., 1988 'Environmental Impact Assessment: Does It Really Work?' in *Habitat Australia*, October, 1988, pp.22-3.

Russell M. & Gruber M. 1987, 'Risk Assessment in Environmental Policy-Making' in *Science*, April 1987, 282-290.

Todhunter J.A., 1986, 'Societal Considerations in Implementing Risk Management Decisions: Towards Improving the Process' in *The Science of the Total Environment*, 51: 63-68. Woodhead 1990:

World Health Organization, 1985, 'Environmental Health Impact Assessment of Urban Development Projects: Guidelines and Recommendations'. June 1985, Geneva.

World Health Organization, 1987, 'Health & Safety Component of Environmental Impact Assessment'. Environmental Health Series no.15, Copenhagen.

World Health Organization, 1988, 'Health and Safety Component of Environmental Impact Assessment'. Environmental Health Series, no. 1, 1-67.